# SPC for Thinkers

Also available from ASQ Quality Press:

*Statistical Quality Control Using Excel, Second Edition*
Steven M. Zimmerman and Marjorie L. Icenogle

*The Desk Reference of Statistical Quality Methods*
Mark L. Crossley

*Glossary and Tables for Statistical Quality Control, Fourth Edition*
ASQ Statistics Division

*The Quality Improvement Handbook*
ASQ Quality Management Division and John E. Bauer, Grace L. Duffy, Russell T. Westcott, Editors

*Root Cause Analysis: Simplified Tools and Techniques*
Bjørn Andersen and Tom Fagerhaug

*The Quality Toolbox, Second Edition*
Nancy R. Tague

*The Path to Profitable Measures: 10 Steps to Feedback That Fuels Performance*
Mark W. Morgan

*Quality Essentials: A Reference Guide from A to Z*
Jack B. Revelle

*The Quality Improvement Glossary*
Donald L. Siebels

To request a complimentary catalog of ASQ Quality Press publications, call 800-248-1946, or visit our Web site at http://qualitypress.asq.org.

# SPC for Right-Brain Thinkers

Process Control for
Non-Statisticians

Lon Roberts

ASQ Quality Press
Milwaukee, Wisconsin

American Society for Quality, Quality Press, Milwaukee 53203
© 2006 by American Society for Quality
All rights reserved. Published 2005
Printed in the United States of America

12 11 10 09 08 07 06 05    5 4 3 2 1

**Library of Congress Cataloging-in-Publication Data**
Roberts, Lon.
  SPC for right-brain thinkers : process control for non-statisticians / Lon Roberts.
     p. cm.
  Includes bibliographical references and index.
  ISBN 0-87389-663-7 (softcover : alk. paper)
  Process control—Statistical methods.   I. Title.

TS156.8.R53 2005
658.5'62—dc22                                                          2005013807


ISBN-13: 978-0-87389-663-4
ISBN-10: 0-87389-663-7

No part of this book may be reproduced in any form or by any means, electronic, mechanical, photocopying, recording, or otherwise, without the prior written permission of the publisher.

Publisher: William A. Tony
Acquisitions Editor: Annemieke Hytinen
Project Editor: Paul O'Mara
Production Administrator: Randall Benson

ASQ Mission: The American Society for Quality advances individual, organizational, and community excellence worldwide through learning, quality improvement, and knowledge exchange.

Attention Bookstores, Wholesalers, Schools, and Corporations: ASQ Quality Press books, videotapes, audiotapes, and software are available at quantity discounts with bulk purchases for business, educational, or instructional use. For information, please contact ASQ Quality Press at 800-248-1946, or write to ASQ Quality Press, P.O. Box 3005, Milwaukee, WI 53201-3005.

To place orders or to request a free copy of the ASQ Quality Press Publications Catalog, including ASQ membership information, call 800-248-1946. Visit our Web site at www.asq.org or http://qualitypress.asq.org.

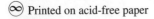 Printed on acid-free paper

Quality Press
600 N. Plankinton Avenue
Milwaukee, Wisconsin 53203
Call toll free 800-248-1946
Fax 414-272-1734
www.asq.org
http://qualitypress.asq.org
http://standardsgroup.asq.org
E-mail: authors@asq.org

*To Sharon—my wife and best friend,
a paragon of right-brain thinkers.*

# Contents

*List of Figures and Tables* ............................... ix
*Preface* ............................................... xi

**Chapter 1**    **SPC in Perspective** ....................... 1
                Options for Dealing with Quality ............. 3
                Management by Fact ........................ 4
                The Six Sigma Approach .................... 4
                The SPC Story ............................. 5

**Chapter 2**    **Framework for Applying SPC** ............... 9
                Setting Up the SPC Subsystem ............... 11

**Chapter 3**    **SPC in Action** ........................... 17
                Case Involving Advocate General ............ 17
                Lessons Learned ........................... 21

**Chapter 4**    **Using SPC to Identify Problems** ............. 23
                Interpreting SPC Chart Patterns .............. 26
                The Cumulative Sum Chart .................. 34
                Characteristics of a Normal Distribution ........ 37

**Chapter 5**    **Analyzing the Process Aspects of Your Job** .... 39
                The Critical Human Factor .................. 39
                Mapping the Process ....................... 42

**Chapter 6**    **Deciding Where to Focus Your Effort** ........ 45
                Brainstorming ............................. 45
                Fishbone Diagram ......................... 46
                Pareto Analysis ............................ 49

**Chapter 7**    **Barebones Statistics** ....................... 53
                Averages and Means ....................... 53

|  | Range Values | 55 |
|---|---|---|
|  | Standard Deviation | 56 |
|  | Populations and Samples | 58 |
|  | Normal Distribution | 60 |
|  | Statistics and Parameters | 61 |
| **Chapter 8** | **Planning Your Charting and Measurement System** | **63** |
|  | The X-Bar and R-Charts | 64 |
|  | Charting Individual Values | 67 |
|  | Monitoring Attribute Data | 68 |
| **Chapter 9** | **Process Capability and Control Limits** | **73** |
| **Chapter 10** | **Collecting and Plotting Your Data** | **77** |
|  | Where and How Should Data Be Collected? | 78 |
|  | Who Should Be Responsible? | 79 |
|  | How Often Should Data Be Collected? | 80 |
|  | Who Inspects or Reviews the Results? | 81 |
|  | Who Takes Corrective Action? | 82 |
| **Chapter 11** | **Continuous Improvement** | **85** |
| **Concluding Remarks** | | **91** |
| **Appendix A** | **Constructing a Pareto Analysis Chart** | **93** |
| **Appendix B** | **Charting Individual Values** | **95** |
| *Glossary* | | *99* |
| *References* | | *105* |
| *Index* | | *107* |

# Figures and Tables

## FIGURES

| | | |
|---|---|---|
| Figure 2.1 | The SPC subsystem. | 11 |
| Figure 2.2 | Setting up the SPC subsystem. | 12 |
| Figure 2.3 | Control chart format. | 15 |
| Figure 3.1 | Patent application process (example). | 19 |
| Figure 4.1 | Control chart with scales. | 25 |
| Figure 4.2 | Control chart showing normal variation. | 27 |
| Figure 4.3 | Sudden out-of-control condition. | 28 |
| Figure 4.4 | Pattern shift. | 30 |
| Figure 4.5 | Cycling. | 30 |
| Figure 4.6 | Trending condition. | 31 |
| Figure 4.7 | Stratification pattern. | 32 |
| Figure 4.8 | Mixture pattern. | 33 |
| Figure 4.9 | CuSum chart. | 36 |
| Figure 4.10 | Stability and trending conditions. | 37 |
| Figure 5.1 | Process flowchart example. | 43 |
| Figure 6.1 | Fishbone diagram. | 47 |
| Figure 6.2 | Process classification format. | 49 |
| Figure 6.3 | Pareto analysis example. | 50 |
| Figure 7.1 | Normal distribution example. | 61 |
| Figure 8.1 | X-bar and R-chart example. | 65 |
| Figure 10.1 | C-chart with comments added. | 83 |
| Figure 11.1 | Before and after improvements. | 86 |
| Figure B.1 | Chart of individuals. | 97 |

# TABLES

| | | |
|---|---|---|
| Table 2.1 | Possible process performance indicators. | 13 |
| Table 4.1 | Data used in preparing CuSum chart. | 36 |
| Table 4.2 | Distribution of variable data. | 38 |
| Table 8.1 | Comparison of sample statistics. | 66 |
| Table 8.2 | Selecting an attribute control chart. | 70 |
| Table 8.3 | Formulas for UCL and LCL on attribute charts. | 72 |
| Table 9.1 | Specification limits versus control limits. | 75 |
| Table A.1 | Data used in constructing the Pareto analysis chart. | 94 |
| Table B.1 | Calculating the moving range and centerline values. | 96 |

# Preface

*SPC for Right-Brain Thinkers* is not simply another made-easy book on the subject of statistical process control (SPC). Though several books on the market do an excellent job of simplifying the subject matter—most notably by going easy on the math and statistical theory—the guiding principle in writing this book was to make SPC accessible to that large group of individuals who would readily characterize themselves as *right-brain thinkers.* As they themselves would be quick to tell you, right-brain thinkers do not reason in a linear-logical fashion—an assertion that is backed by research conducted by the Nobel Prize–winning psychobiologist, Roger Sperry. The challenge that right-brained thinkers face in understanding and applying SPC goes beyond the math, though the math may indeed be a barrier. It is also a matter of approaching the subject from a different perspective altogether—through the side door, if you will, where the inner workings of SPC may be seen in action.

While this book will appeal to right-brain thinkers, it is also intended to serve the information needs of those who either *own* or *work within* the job processes wherein SPC is applied. These are intelligent individuals who have a stake in the day-to-day management and support of these processes, often in the capacity of front-line problem solvers or contributors to process improvement endeavors. It is not, however, a technical how-to book for specialists who are responsible for the details of implementing a newly minted SPC tracking system. This base is well covered by any of the many expanded textbooks on statistical process control, some of which are listed in the References.

In keeping with the intent of making this book's structure as well as its contents accessible to the target audience, the order in which the material is presented may appear somewhat nontraditional. Also, since service industries employ 80 percent of all U.S. workers—and right-brain thinkers are inclined to gravitate to service-oriented jobs—the examples used in this book demonstrate the use of SPC in a service organization: an imaginary law firm called Advocate General. While any book that seeks to instruct will require some degree of adaptation on the part of the reader, the examples used here demonstrate the basic principles without requiring specialized knowledge of a particular industry, legal or otherwise. Also, the basic framework for deploying an SPC subsystem is embedded within a case scenario that can be adapted to a variety of industries in the able hands of those who have specialized knowledge of their own internal processes. The basic format and interpretation of a control chart used to track response times in handling customer requests is the same, for instance, regardless of whether the customer request is initiated by a stock purchaser or a patient in a hospital emergency room.

Here, then, is a composite profile of the individuals who will likely be attracted to this book:

- Those who are inclined to label themselves as right-brain thinkers
- Those who are intimidated by math, possibly even the mere mention of something as ominous sounding as "statistical process control"
- Those who need only a basic understanding of SPC, perhaps from a systems perspective or as a potential user of an SPC tracking system

Although individuals who match this profile can be found in virtually every walk of life, here are some circumstances where they and this book may intersect:

- In an academic program that surveys a variety of process management tools and methods, such as a quality course for non–engineering majors
- In a corporate training program geared for process stakeholders, such as decision makers or individual contributors
- In the briefcase of a business manager or process owner who is interested in a quick read on SPC, perhaps during a cross-country flight
- In the information toolkit of those who are actively involved in process improvement initiatives, perhaps as a member of a process improvement team

Finally, for those who are interested, a variety of articles and spreadsheet templates on statistical process control, and related topics, can be accessed from the author's website, www.R2assoc.com.

May this book be instrumental in helping you discover the power and benefits of SPC, giving those who seek it an alternative "door" to access this important tool.

—Lon Roberts

# 1 SPC in Perspective

> A complex system that works is invariably found to have evolved from a simple system that worked.
>
> —John Gall
> *Systemantics*

Not that I have been asked to do so, but if I were given the task of assigning a label to what is referred to as "statistical process control," I would reorder the terms to ensure the emphasis would be given where it belongs. Though less succinct than statistical process control, something on the order of "process control aided by statistical techniques" would put the emphasis where it belongs: on the intent (process control) rather than the tool (statistical techniques).

Throughout my career of training and consulting others, I have encountered many who are reluctant to use statistical process control—SPC for short—because they do not understand how and where it may be applied. Some are so turned off or intimidated by the "statistical" part of the label that they simply avoid the subject altogether. Far too many others, I am sad to say, fail to recognize the "process aspects" of their workplace, let alone the fact that such processes are worth the effort of being controlled through the use of statistical techniques in the first place.

Perhaps the reason for all of this stems from the environment in which SPC has been most closely aligned. There was a time when SPC was limited almost exclusively to the realm of manufacturing processes—specifically those involved in mass production. As a rite of passage, anyone who aspired to understand SPC was first exposed to a heavy dose of statistical theory, perhaps because of the dominant thinking style of "left-brained types" who have traditionally been the purveyors of SPC knowledge.

Fortunately, the situation has changed over the years as the benefits of SPC have become much heralded and convincingly clear. As it stands today, SPC is being used well beyond the manufacturing environment. It is not unusual, for instance, to find workers using SPC in such diverse industries as health care, transportation, and banking. What's more, SPC has moved beyond the realm of being used exclusively to control product quality. Many companies have found SPC to be an effective tool for controlling quality performance in the delivery of services as well. Even manufacturing companies are finding applications for SPC beyond its traditional use in monitoring and controlling production line quality.

In conjunction with the now broader interest in SPC—in other words, beyond the realm of product manufacturing—many companies have also moved toward *decentralizing* the responsibility for quality. Workers throughout the organization are being told that they have a direct role in controlling the quality of the services or products ultimately delivered to the customer. As a result, SPC is being widely used to monitor and control processes that, to some degree, concern everyone in the organization, regardless of their job description or educational background.

Furthermore, in part because of the excitement generated by Robert Kaplan and David Norton over something called the *balanced scorecard*, there is heightened interest in the use of meaningful business metrics that are beneficial for "translating an organization's mission and strategy into a comprehensive set of performance measures that provides the framework for a strategic measurement and management system" (Kaplan and Norton, 1996, p. 2).

The balanced scorecard advocates monitoring key metrics that are essentially *drivers of future performance*. These metrics are drawn from four sources: financial data, customers, internal processes, and indicators of learning and innovation.

This emphasis on proactive indicators of performance has much in common with SPC. Indeed, SPC may be used to monitor and track the variables that are considered the performance drivers within a comprehensive balanced scorecard system. In contrast to "dashboard indicators" that provide point-in-time indication of how these business metrics are performing, SPC charts offer the added benefit of displaying how these metrics are performing over time and discerning whether or not a certain pattern represents a trend—important factors to consider if *anticipation of future performance* is a primary motivation for using a balanced scorecard system in the first place.

## OPTIONS FOR DEALING WITH QUALITY

Companies employ a variety of ways and means for dealing with the quality as it pertains to their products or services. The basic options consist of one, or more likely a combination, of the following:

1. Simply ignoring the whole notion of quality and dealing with the consequences later on the customer end if necessary
2. Thoroughly inspecting the product or service just before "handing it over" to the customer, and then incorporating any fixes that may be necessary
3. Monitoring certain "early warning indicators" that tell them something significant about the process itself while the product or service is still in production
4. Conducting a series of "experiments" to garner meaningful data that may be used in optimizing the process from the outset, allowing quality to be designed into the process
5. Making continuous improvements to such a degree that the whole matter of having to pay attention to quality eventually becomes a moot point

In its purest sense, statistical process control is most closely aligned with the third category—taking proactive measures to ensure that the process is behaving as it should *before* the end product or service is actually produced. But, it may also be employed in conjunction with either of the latter two categories.

For instance, in the context of category 4, once the design of a certain process has been optimized based on the outcome of the controlled experiments, SPC might be used to monitor those factors (or variables) that were identified by the experiments as critical to optimum performance. In the context of category 5, SPC will likely play a significant role in monitoring process performance and identifying improvement opportunities—at least until the desired nirvana state is achieved.

Before leaving this subject, it is worth noting that the inspection method, suggested by category 2, may indeed be effective at preventing quality problems from reaching the customer, but it tends to be unnecessarily costly and time consuming, except where extremely critical situations warrant its use. Not only does it slow down delivery of the product or service, but it also reduces the "yield" of the process by forcing a reaction to problems at or near the end of the process rather than *during* the process. Also, if the number of individual "rejects" reaches a certain predetermined level, complete lots or batches may have to be rejected in order to reduce either the buyer's or the seller's risk. Statistical techniques that pertain to this form of reactive quality management are often labeled as *statistical quality control,* as opposed to *statistical process control.*

## MANAGEMENT BY FACT

Since the *output* of SPC consists of objective, fact-based information on how a certain process is performing, statistical process control is an important element in the contemporary management philosophy referred to as "management by fact." In a nutshell, management by fact, or MBF for short, is an approach to managing people and processes that is predicated on making decisions that are *informed* by factual information, rather than groupthink, guesswork, or wishful thinking. MBF is made possible by *listening* to the "voice of the process"—in other words, timely and meaningful measurement data taken from the process in question.

In the ongoing war against waste and inefficiency, MBF is an especially powerful weapon in the arsenal of an experienced manager—one whose knowledge and already keen ability in exercising judgment is further enhanced by facts and careful observation.

## THE SIX SIGMA APPROACH

The Six Sigma approach to quality management is a contemporary strategy that gives substance to our category 5 classification—that is, making continuous improvements to the extent that tracking and monitoring quality performance essentially becomes a nonissue.

The Six Sigma school of thought had its genesis at Motorola in 1979 and is predicated on research by the late Bill Smith, who was a Motorola engineer at the time. Looking for ways to reduce waste, Smith discovered that if a product found to be defective is corrected *while it is still in the production phase,* that product is more likely than otherwise to contain *additional* defects that go undetected until the product is in the hands of the customer. Conversely, Smith found, products produced in an error-free environment are less likely than otherwise to experience problems later.

Fueled by this study, the initial intent of the Six Sigma approach was to make continuous improvements in the process in question to such an extent that fewer than 3.4 defects, on average, would occur out of a million opportunities. If, for example, you were to write a book consisting of exactly one million words and were to discover only three misspelled words in the entire manuscript, you would be within Six Sigma guidelines—at least for this one particular characteristic: *properly spelled words*. Since there are precisely one million *opportunities* for a spelling error to occur in this example, it can be said that we have achieved a quality level of three defects per million opportunities, or in shorthand, three DPMO. Also, in the language of the Six Sigma approach, we could say that *properly spelled words* is a requirement that is *critical to quality* (CTQ), if this is indeed the case.

Before leaving this subject, it is worth noting that while the outcome of Smith's study initially resulted in a quest to ferret out and essentially eliminate defects by modifying the process in question, the domain of Six Sigma has expanded over the years. According to a 2003 brochure distributed by Motorola University: "We have evolved Six Sigma beyond a manufacturing approach for counting defects to a strategic methodology that applies to all business functions."

Furthermore, in cautioning that the Six Sigma calculations are not well suited to "human intensive processes," another Motorola employee writes: "In the case of human resources, the definition of a defect, such as employee performance that falls below a certain level, can be controversial and can also be manipulated to get a better sigma value" (Barney, 2002, p. 15).

Consequently, the current incarnation of Six Sigma—at least from Motorola's perspective—focuses more on applying Motorola's "strategic methodology" than on reducing defects to less than 3.4 DPMO. Nevertheless, in certain circles, there remains a lively debate about the precise statistical interpretation of Six Sigma as a metric.

## THE SPC STORY

Quality management philosophies have fallen in and out of favor over the years, due in part to the evolution of knowledge and due occasionally to the influence of certain visionary leaders, though often, I would submit, as a consequence of our human propensity to simply embrace newness—the latter a characteristic that methodology consultants understand well. In contrast to complex quality management systems that wax and wane, the staying power of statistical process control, with its simple elegance, is testament of its proven worth as the focal point around which a successful comprehensive quality management system can be built. In an article in the July 2003 issue of *Quality Progress* magazine, author Lynne Hare expressed a similar point of view:

> SPC grew into everything we do. It changed the way we think, work and act, and it evolved into total quality management (TQM). But there is an ebb and flow to new technologies in our society, and TQM's star faded in the presence of reengineering, which faded with the advent of Six Sigma. All the while, the basic SPC tools have been refined and augmented, and SPC serves in muted presence to underpin the newer, expanded technologies. (Hare, 2003, p. 63)

From its genesis in the 1920s in the mind of Walter Shewhart (1891–1967), who was working in the research labs of the Western Electric Manufacturing plant in Hawthorne, Illinois, until now, statistical process control has enjoyed a track record of success over 80 years. Though other tools have been added to the basic "toolkit" over the years, the fundamental component of any SPC system is the *control chart*—a simple chart that plots the quantitative value of a certain quality characteristic over time, with the intent of monitoring its pattern of variation and making rational, fact-based decisions on whether the underlying process is behaving normally or abnormally. As noted earlier, although SPC originated and was honed in a manufacturing environment, over the past several decades it has been used in a variety of service organizations as well, both in the public and private sectors.

Before we examine SPC in action, we should take note of certain factors and conditions that inherently make SPC a valuable tool. These may be summarized as follows:

1. **Management support:** SPC is a strategic tool. Senior management must take an active role in ensuring that SPC will be deployed and maintained and that the resulting data will be used to make fact-based decisions that affect the structure and operation of the process or processes in which it is applied.

2. **Process orientation:** SPC forces a clear definition of the elements of the process, including where the process begins and ends, how its elements interact, and where the process fits in the system as a whole. It also requires the identification of critical points within the process that are prime candidates for being monitored and controlled with the aid of SPC.

3. **Problem prevention orientation:** SPC works inside the process to ensure that potential problems are detected and headed off *before* they culminate in the product or service that reaches the customer. Problem avoidance not only saves time and money, but it also promotes good will with the customer.

4. **Voice of the process data:** SPC tracks quantitative data that are based on objective measurements or other observations from within the process. Furthermore, the control limits used in constructing the various SPC control charts are predicated on data collected from the process so that normal behavior can be distinguished from abnormal behavior over time.

5. **Understanding and controlling variation:** Any time a product or service specification is not met, the organization—and perhaps the end customer—will be affected in a negative way. Therefore, SPC takes a special interest in understanding and controlling process variation. The SPC control charts provide the means for monitoring process variation.

The significance of these five factors and conditions will become more evident as we explore the SPC framework in the next chapter and the case scenario that follows it.

# 2 Framework for Applying SPC

When using the term *process control* in a quality-performance context, we are referring to the action involved in monitoring and controlling a certain process to ensure it *consistently* stays within an acceptable range of specified limits. In the language of SPC, these limits are referred to as *control limits*. By modifying the term *process control* to read *statistical process control,* we enhance the meaning to suggest a basic change in the way the process is monitored. Let's see how.

When employing statistical process control, it is typically our intent to make only occasional measurements, rather than continuous measurements, of the performance factors that we are interested in monitoring. Again, using the language of SPC, we refer to each of these occasional measurements as a *sample*. The factors we choose to measure are referred to as *variables* for the simple reason that even under normal circumstances we might expect to see a slight degree of variation in the results between samples. This variation between sample results, however slight it may be, could be due to a number of causes, which we will examine later.

> **Exercise:** *Briefly describe a process in which occasional measurements of performance factors might be preferred over continuous monitoring. (Note: Any process that involves large quantities of data is a likely candidate.)*
>
> _____
> _____
> _____
> _____

In any case, whether we choose to occasionally sample or to continuously monitor, the performance factors themselves should provide a direct, or strong indirect, indication of an actual or potential problem in the original process.

> **Exercise:** *Describe one or more situations in which indirect measures of quality performance might be preferred or even necessary. (For example, in measuring customer satisfaction, it may be preferable to monitor certain purchasing habits rather than ask for subjective ratings from the customer.)*
>
> _____
>
> _____
>
> _____
>
> _____

When a problem does occur, at least as suggested by one or more of our samples, we say that the process is *out of control*. SPC certainly will not fix or even prevent the actual process problem, but it should provide us with reliable information upon which to base corrective, or preferably, preventive action.

Quite often we are interested in a pattern, or a trend, rather than the results of a single sample. Using SPC to establish trends is an excellent way of heading off problems before they become too severe, or perhaps irreversible.

It may be helpful to think of SPC as an element, or subsystem, of some larger system, as depicted by Figure 2.1. As this illustration shows, the purpose of the process is to accept inputs and transform them into outputs. In an insurance firm, for example, the inputs for a certain process may include claims, verification of coverage, and accident reports, while the outputs may include correspondence, payment of claims, and updated client records.

The SPC subsystem monitors one or more *control points* (or quality characteristics) within the process. Once the measurements are *charted* and interpreted, corrective or preventive action can be prescribed to bring, or keep, the process in control.

It is important to stress that SPC is primarily interested in what goes on within the process *before* the outputs are produced. While output measurements may be all we have to work with in some cases, they are typically less helpful in avoiding problems than measurements taken within the process. Thus, one of the basic reasons for

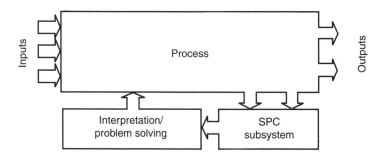

**Figure 2.1** The SPC subsystem.

using SPC in the first place is to keep the process in check before problems become manifest at the output stage—or at least before they are irreversibly out of control. There are two important reasons behind this:

1. Once a problem reaches the output level, there is a good chance it will be passed on to the customer. Quality suffers—often beyond repair—anytime the customer is on the receiving end of the problem.

2. It is less costly to fix a problem upstream rather than downstream. In other words, the further along we are in the process before detecting a problem, the more it will cost, in time and money, to correct.

Before leaving Figure 2.1, we should stress the point that the outputs of the process may be tangible or intangible (or both). Tangible outputs most often take the form of products, services, or information. While intangible outputs are a little more difficult to nail down, they are important all the same. Intangible outputs may take the form of customer relations, congenial service, or being responsive, to name a few.

## SETTING UP THE SPC SUBSYSTEM

Figure 2.2 depicts the primary tasks involved in setting up the SPC subsystem—the basic framework. In Chapter 3 we will apply these tasks to our case scenario to witness the framework in action. First, let's briefly examine each of the major tasks in descriptive terms.

## 12 Chapter Two

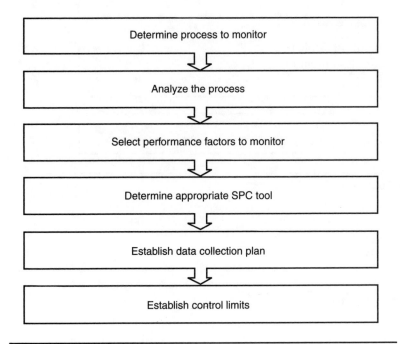

**Figure 2.2** Setting up the SPC subsystem.

### Determine Process to Monitor

First, it may be necessary to give some thought to which aspects of the overall system we wish to monitor and control by using SPC. Though there may be other considerations, those processes that have the greatest impact on quality and customer satisfaction are prime candidates. Even though its roots are in the product manufacturing environment, SPC can be adapted to virtually any process. The fact that it has long been used in the product-manufacturing environment simply attests to its value, rather than to a lack of versatility.

### Analyze the Process

Next, the selected processes are analyzed in terms of their own sub-processes. Inputs, outputs, and the action fulfilled in transforming the inputs into outputs are determined and perhaps diagrammed. Interface points between the various processes should also be considered.

## Select Performance Factors to Monitor

It is now possible to select the performance factors, or indicators, that we wish to monitor and measure. This is a critical step in setting up the SPC subsystem, one that often requires us to return to the previous step (process analysis) to ensure we understand exactly what it is we are attempting to measure and why. If the wrong parameters are monitored, the results will be of little value in helping us control the process. At the very least, we waste valuable time in collecting the wrong data. Since, for practical reasons, we cannot monitor everything that could possibly go wrong, we are particularly interested in *critical control points*—those points within the process where a small change can have a relatively large impact on the output.

The specific factors that we choose to monitor will be unique to the process being considered. Table 2.1 provides some examples of

**Table 2.1** Possible process performance indicators.

| Function | Performance indicator |
|---|---|
| Shipping and receiving | • Bill of lading errors<br>• Routing errors<br>• Internal notification time |
| Purchasing | • Purchase order errors<br>• Order processing time<br>• Purchase rejects |
| Sales | • Quotation errors<br>• Sales order errors<br>• Inquiry response time |
| Engineering | • Design change notices<br>• Specification errors<br>• Calculation errors |
| Customer service | • Inquiry response times<br>• Telephone etiquette ratings<br>• Number of return calls<br>• Customer satisfaction ratings |
| Information systems | • System downtime per install<br>• Coding errors per module<br>• Frequency of user interface errors |
| Production | • Percent defective items<br>• Warranty recalls<br>• In-process conformance measurements<br>• Wait times (at various points in the process) |

generic parameters that may be monitored in relation to certain functions within the organization.

In reviewing performance factors that you may wish to monitor, keep in mind that it is desirable to base your selection on those that will provide *early indication* of an impending problem. Depending on the process, some of the indicators listed in Table 2.1 meet this criterion better than others. If intermediate control points are not immediately obvious, it may be necessary to define the process in greater detail to identify points where potential problems can be identified and prevented further upstream.

Notice also from Table 2.1 that certain performance factors are quantitative in nature and others are qualitative. Quantitative measures are based largely on objective facts, and qualitative measures are often based on perceptions. Again, depending on the process, this does not suggest that one is superior to the other. In service-oriented jobs, in particular, it is the customer's subjective *perception* of reality that ultimately matters the most.

### Determine Appropriate SPC Tool

After selecting the performance factors to monitor when setting up our SPC subsystem, we must next determine the most appropriate SPC tools and techniques. While we will have more to say about this later, suffice it to say that certain SPC tools are designed for use with measurements that can assume virtually any numerical value (within a range), and others are designed to be used in situations in which the values can assume only one of two conditions—such as good versus bad, error versus non-error, or operable versus inoperable.

### Establish Data Collection Plan

Our next step is to establish a data collection plan. The data collection plan should address how, where, and when the measurements will be taken. Such a plan should also specify who is responsible for making and recording the measurements, including who is responsible for taking corrective action when it becomes necessary.

### Establish Control Limits

The final step in setting up the SPC subsystem concerns the establishment of control limits, as depicted in Figure 2.3, to indicate the upper and lower boundaries for process performance under normal conditions. These control limits are predicated on data gathered from the process, using basic statistical techniques to determine values for the upper and lower control limits—hence the name statistical process control. Furthermore, they are spread apart at a distance that has shown to strike a balance between letting a problem go unnoticed and being overreactive to normal process fluctuations. It is important to recognize that the use of fact-based control limits takes the guesswork and subjectivity out of trying to determine whether a process is performing within the bounds of its capability.

Collecting data to establish the control limits is accomplished by observing the process and measuring the control points we are interested in over a period when there are no known problems. In other words, we are establishing the normal operating conditions, allowing for the fact that a certain degree of variation is acceptable and even to be expected. If, after establishing the control limits, we sample a certain performance factor and find it to be outside either of these control limits, corrective action is called for. Preventive action may likewise be called for if the process is *trending* toward an out-of-control condition or if the control chart indicates a pattern of behavior that is not likely to be caused by normal fluctuations, or what statisticians refer to as an expected pattern of *random variation*.

It should also be noted that the *average* level of performance is determined in this step as well. In fact, as we will later see, the average value is actually used in establishing the upper and lower control limits, and it lies in the center between these two control limits. Furthermore, the average value typically defines the level around which we monitor long-term variations in process performance, where *long-term* typically refers to a succession of samples rather than some distinct period.

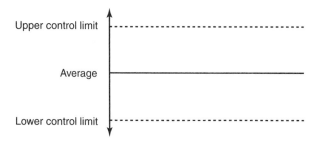

**Figure 2.3** Control chart format.

A couple of points should be stressed before we leave this chapter. First, what Figure 2.2 cannot show, but which are important to consider nonetheless, are the human and political realities within the organization that will either make or break the SPC subsystem. Not only is it necessary for people to believe that SPC can help them improve their processes, but they should also be given an active part in planning and making it work. While for some processes the preparation of SPC control charts is automated—as with certain manufacturing processes—in any case there is no substitute for human judgment when it comes to interpreting the charts and prescribing corrective action where needed. Some would also argue that maintaining the control charts by hand encourages "thoughtful intervention" on the part of the people who support the process in question. While this may be impractical with mass production processes, where tens of thousands of data points can be generated over a short time, it is certainly feasible for most service-oriented processes, where the data stream is typically more manageable.

We should also stress the fact that SPC does not operate in a vacuum. Once our SPC subsystem is in place, it becomes one of possibly several important tools for monitoring and controlling the performance of the process. If it is desirable to *tighten* the control limits of the process, for example, to be more competitive or more customer-focused or to diminish waste over the long run, then action must be taken to improve the process itself rather than arbitrarily manipulating the control limits. SPC cannot help in this regard—it can tell us only if the process, as it exists, is performing in a manner we would expect, given the circumstances.

Now that we have taken a big-picture look at the framework for implementing an SPC subsystem, in the next chapter we will introduce the human element into this process by examining an application scenario.

# 3 | SPC in Action

Those of us who view ourselves as right-brain thinkers know that we learn concepts best by seeing applications in action rather than by laboring over theories, formulas, and wordy descriptions. With this in mind, this chapter looks briefly at a scenario involving the application of SPC to set the stage for the concepts that are discussed later. We will do this by examining a case pertaining to an imaginary law firm called Advocate General.

The case has been greatly simplified to avoid any unnecessary difficulty in understanding the context for the application—you don't need to be a lawyer to comprehend the application, at least at the level of detail it is presented here. Certain aspects of this application will be expanded on in later chapters as the SPC principles are introduced.

## CASE INVOLVING ADVOCATE GENERAL

Advocate General is in the business of providing legal services to mostly small corporations that do not have a legal staff of their own. In addition to providing general counsel, the firm is divided into three areas of specialization: contracts, labor law, and intellectual property. Overall, the firm employs 54 people, consisting of 5 partners, 30 staff lawyers, 10 clerks, 5 administrative assistants, and 4 legal assistants.

In response to the need to become more customer focused, the firm's partners have decided that their three primary specialties should be operated as processes, with appropriate checkpoints and controls to ensure that all work details are accomplished in a timely fashion and to the satisfaction of the client. It has also been determined that SPC should be used to monitor and control these processes.

One of the partners, Jean Sanders, has been chosen to coordinate the implementation of this new system because of her seniority and her familiarity with SPC from having worked on a quality planning task force while serving on the legal staff for a civil engineering firm. Following the task sequence in Figure 2.2 (see Chapter 2), here is a summary of what was accomplished.

### Determine Process to Monitor

It was determined that each of the three specialties—contracts, labor law, and intellectual property—should be operated as a process. Upon further examination it was determined that each specialty could be classified into certain processes that were interactive but somewhat distinct when viewed as end-to-end processes that accepted inputs and provided outputs. (We will focus on one of these: patent application process.)

### Analyze the Process

Ms. Sanders chose senior attorneys from each specialty to serve as the primary representatives for their respective areas. These representatives met with the entire staff from their specialties to brief them on the decisions that had been made up to this point and to give them a role in analyzing the processes within their specialty.

To assist in the analysis, each team constructed a process flowchart showing the major tasks within the process, as well as the interrelationship between these tasks. As the project leader, Ms. Sanders demonstrated how the process flowchart could be beneficial in identifying potential trouble spots—such as bottlenecks—as well as critical points that may need to be monitored, perhaps with the aid of SPC.

Figure 3.1 illustrates the primary tasks and sequence of tasks that were determined by attorney Bob Taylor's team to make up the patent application process. This particular process falls under the intellectual property specialization. It was also determined from historical data that the patent application process handles an average of 500 applications per year.

### Select Performance Factors to Monitor

Mr. Taylor's team determined that a number of control points in the patent application process could be monitored. From this list, it was

# SPC in Action 19

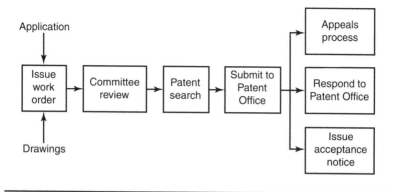

**Figure 3.1** Patent application process (example).

determined that only three critical control points needed to be monitored by using SPC. These were:

- Number of mistakes on the application form
- Percentage of initial rejections from the Patent Office
- Delay in passing the application through the Review Committee

## Determine Appropriate SPC Tools/Techniques

Based on the three critical control points, it was determined the following SPC tools should be applied:

1. *Number of Mistakes on the Application Form* requires a count of the number of mistakes that are found on the standard application form. From this information, it was determined that a C-chart was the most appropriate SPC tool (we will later discuss why).

2. *Percentage of Initial Rejections from the Patent Office* requires the simple calculation of the number of rejections divided by the total number of submissions over a certain period of time. Using this information, it was determined that the P-chart was the most appropriate SPC tool (again, we will later discuss why).

3. *Delay in Passing the Application through the Review Committee* requires a measurement of time. Because time can be measured on a scale ranging from zero to infinity, it was determined that the X-bar and R-chart were the most appropriate SPC tools (we will later discuss why).

The previously discussed charts and other SPC charting tools are expanded on in Chapter 8.

## Establish a Data Collection Plan

Mr. Taylor's team determined that the following factors should be considered in establishing a data collection plan in relation to each of the critical control points:

- How the data would be collected
- How often data would be collected
- Who would be responsible for collecting the data
- Who would be responsible for interpreting and acting on the results

The results of this plan are shown as follows for the Mistakes on Application Form control point:

**How collected:** The firm's standard application form will be periodically reviewed for transposition errors, missing data, factual errors, and inconsistencies by tabulating a count of all discrepancies.

**How often:** One out of every three patent applications will be examined for discrepancies. With 500 applications per year, this amounts to an average of 3.3 applications per week. Time is to be allocated accordingly.

**Data collection:** The task of collecting data should be performed on an alternating basis by the administrative assistant and the legal assistant assigned to the department.

**Action:** The team determined that the department administrative assistant should be given primary responsibility for monitoring the results and keeping this aspect of the process in control. The legal assistant will back up the department administrative assistant, and the charts are to be reviewed by a clerk.

### Establish Control Limits

Mr. Taylor's team determined that historical data can be used to establish the average values, as well as the control limits, for each of the three control charts. To use these data, it will be necessary to eliminate any values that correspond to past problems that can be assigned to known causes. In other words, in establishing the average values and the control limits, Mr. Taylor's team will use only the data that are known to represent normal conditions. (For example, the team elected not to include a time measurement for an unusually long Delay in Review that was known to have been caused by a certain committee member's absence during a period of extended illness.) The team also acknowledged that these values—that is, the averages and the control limits—would need to be adjusted if the process is later improved.

## LESSONS LEARNED

Let's reflect on several important points that this example highlights before we proceed. The gaps will be filled in later.

First, we saw how the planning sequence shown in Figure 2.2 could be used to set up an SPC subsystem within a process. In this particular case, the process was distinctly different from that found in a product-manufacturing environment. It is naturally easier to visualize how SPC works in a manufacturing environment, where measurements involving an array of sophisticated instruments are constantly being taken. Nevertheless, SPC is equally adaptable to service and support processes when such processes are analyzed in terms of inputs, subprocesses, outputs, and control points.

Also, this example at least hints at the human and political undertones that are inherent in setting up and deploying any SPC subsystem. While some of these factors are a function of the unique culture of an organization, the following considerations are common management concerns in many, if not most, circumstances.

When SPC is introduced, there will undoubtedly be those who feel that their job is in jeopardy. Resentment may turn into outright rebellion if SPC is billed as a tool for cutting waste and inefficiency rather than as a tool to aid in solving problems and enhancing customer satisfaction. As a tool, it is more accurate to think of SPC as a microscope rather than as a hammer or a hatchet.

Care is also necessary to ensure that SPC does not in fact, or in perception, add to the workload of those who are already putting forth a supreme effort to get their jobs done. It should be clear to everyone involved just how much effort will be required to collect data, maintain the control charts, and initiate corrective or preventive action. The person, or persons, who will ultimately be responsible for these tasks should be involved in planning their own roles and responsibilities. Still, some of these concerns may linger until the SPC subsystem has had a chance to demonstrate its value. Some may even find that SPC forces problems to the surface that have long been the source of frustration.

As was pointed out earlier, the very term *statistical process control* intimidates many people. We hope to demonstrate that SPC requires the need to understand little more than a few facts concerning statistics. In any case, no special knowledge of statistics is required to monitor the control points, record data on SPC charts, and detect actual or potential problems by using the charts. Furthermore, no degree of statistical proficiency can substitute for understanding the process itself. (This is where the people working directly with the process have a distinct advantage, which is a significant argument for making SPC accessible to everyone involved with the process.)

Finally, there may be those who initially feel that the SPC way of doing things does not fit within their job description. This is understandable if the message is not clearly communicated that SPC is a tool for doing one's job better, rather than expanding one's current job responsibilities. In this regard, there is no substitute for a top-down commitment to SPC throughout the organization. In fact, when properly used, SPC becomes an excellent communications tool that bridges over many of the common barriers that arise from the hierarchy and functional segmentation of the organization. When an organization begins to see itself as a set of interlinking processes, where the focus is on customer satisfaction—a focus that can become sharper with SPC and its emphasis on *management by fact*—many of these barriers become exposed to the light of day and are eliminated as impediments to progress.

In the next chapter we will examine some of the more common control chart patterns and discuss what these patterns reveal about the process characteristic in question.

# 4 Using SPC to Identify Problems

There would be little need for SPC if we were able to keep a process running smoothly from the moment when it is first put into place. This, of course, is not possible. Minor variations in the process will always be present. Furthermore, every process is likely to have major problems from time to time. Major problems can emerge gradually, as with normal deterioration, or instantaneously, as when some aspect of the process breaks down at what often seems to be the most inconvenient time.

With processes involving machines, wear and tear on the machines will eventually lead to other problems if left unchecked. Processes involving people are subject to variation as well, because of a wide range of factors such as fatigue, miscommunication from point to point or person to person, differences in individual abilities, variances in work ethic, and even mood swings. In the more perplexing cases, a number of factors combine to form complex problems.

In the language of SPC, the term *special causes* refers to problem causes that occur out of the ordinary. In other words, whenever a process goes out of control, we know that special circumstances led to that condition. In common language, the term *firefighting* often describes our reaction to such problems. Special causes are also referred to as *assignable causes,* since they point to some identifiable, though perhaps obscure, cause.

On the other hand, we use the term *common causes* to refer to causes of variation that are inherent in the process itself. Since common causes of variation occur randomly, they are harder to pinpoint. To reduce variation in the process as a result of common causes could require a major overhaul, rather than restoration, of the process. For this reason, the organization's management team is often cited as having primary responsibility for initiating improvements in long-term process performance to reduce the inherent variability stemming from common causes. It is ultimately such improvements that have the greatest influence on competitiveness and on acquiring the ability to exceed the customers' expectations. We will have more to say about common causes of variation when we later discuss the notion of continuous improvement.

While SPC can be used to visually highlight process variation due to common causes, SPC's most important contribution is in spotting problems created by special causes. In this regard, SPC is most effective as a problem-solving tool whenever:

1. It provides a reliable indication of reality.
2. It points to problems as close to the source as possible.
3. Those responsible for its use are able (and committed) to understand and interpret the control charts.

The first issue—concerning reliability—can come into question if, for some reason, the samples taken during our measurements are not representative of the available data in their entirety. Reliability may also suffer if our measurements are persistently inaccurate, that is, if they are off the mark or if they lack precision because of measurement inconsistencies. Precision and accuracy, or the lack thereof, are likely to surface as a concern if special instruments are used in making our SPC data measurements, whether they be mechanical, electronic, or paper-and-pencil assessment devices.

The second issue—concerning proximity to the problem source—is related to two factors: (1) the nature of the data we choose to collect, in other words, whether the data represent variables or attributes; and (2) how well we have analyzed the process in terms of its subprocesses and the factors that influence these subprocesses. If, for instance, we elect to monitor only the final output of a process, the resulting data will be less helpful in pinpointing problems than we might expect from monitoring specific subprocesses upstream from the endpoint. While we will later have more to say about the first of these two factors—that is, the nature of the data—for now we need only point out that variables represent measurable data that can assume virtually any numerical value, typically within a certain range, while attributes represent nonmeasurable data that can only be tested in terms of the presence or absence of a certain quality, such as defects. (Additional details concerning variables and attributes data are addressed in Chapter 8.)

The third issue mentioned above is rooted in the human dimension. No matter how much effort may have gone into the design of an SPC subsystem, it will serve little purpose if users do not know what the results are telling them (as Figure 2.1 suggests). This issue of understanding and interpreting the control charts is the primary focus of the current chapter. With one exception, we will deal with this issue without referring to a specific type of control chart, since the various control charts we later discuss are interpreted similarly when used for problem solving.

Before doing this, take a closer look at the control chart format we saw earlier in Figure 2.3. Notice, in Figure 4.1, the addition of two scales: one along the bottom and another along the left side. The bottom scale denotes instances when data are collected. For example, "6" refers to the sixth data point we intend to record on this chart, perhaps at some specified time after the fifth data point is recorded. The vertical scale on the left represents values we either measure or count. If, for instance, the sixth data point is at 70, this suggests that the variable we are measuring (or attribute we are counting), has a value of 70 on the sixth time it is sampled.

This is a good place to point out that the vertical scale along the left side of the chart will be marked off in increments that correspond to the range of values that the variable or attribute we are monitoring can assume. Here the scale runs from 0 to 100. For another chart it might, for example, run from 0.25 to 0.95.

Notice also the addition of seven values on the vertical scale that are designated as UCL, +2s, +1s, Average, −1s, −2s, and LCL. These values are established in the beginning by observing the performance of the process, over a certain period, during the absence of any problems due to special causes. It may help to think of these as

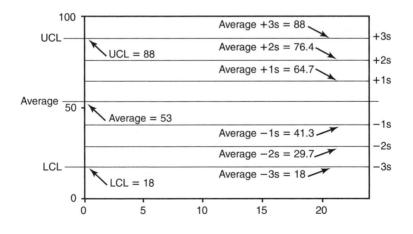

**Figure 4.1** Control chart with scales.

*baseline* values that describe the performance of the process under normal operating conditions.

Also note in Figure 4.1 that the upper control limit is 88, the lower control limit is 18 and the average is 53. Since the scale markings on the vertical scale are graduated in increments of 10, plotting these values (88, 53, and 18) on this particular chart would require some "reading between the lines." In practice it is recommended that the scales be set so that the data points can be plotted relatively close to the horizontal grid lines. This will facilitate plotting the data points as well as reading the charts.

Continuing with this example, notice that the UCL and the LCL are an equal distance from the average (which is 35 increments in this case). The +1s and +2s values simply divide the distance between the average and the UCL into three equally spaced intervals, which in our example are 11.67 increments apart since 35 ÷ 3 is approximately 11.67. The −1s and −2s values do the same for the distance between the average and the LCL. Knowing this, we could have also labeled the UCL as +3s and the LCL as −3s.

Suffice it to say for now that distance between the average and the +1s value refers to something called the *standard deviation*—thus explaining the lowercase "s" designation. Also, the distance between any of the adjacent divisions from the LCL to the UCL are spaced s-units apart. As a result, the LCL and the UCL are each three standard deviations (that is, +3s and −3s) away from the average value. (Additional details concerning the standard deviation and how it is calculated are provided in Chapter 7.)

A final point should be made before leaving Figure 4.1. Our example shows the LCL to have a value of 18. In reality the LCL can go as low as zero in cases where our measurements cannot assume negative values. As a result, you may run across situations in practice where the distance from the average and the UCL (that is, UCL minus average) and the average and LCL (that is, average minus LCL) are not equal if the LCL cannot be pushed below zero.

## INTERPRETING SPC CHART PATTERNS

Now that we have seen the basic format of a control chart, let's see how they can be used to identify problems or potential problems. To begin, let's examine the control chart shown in Figure 4.2. Notice we have used the same values for the average, UCL, and LCL that were used in Figure 4.1, where average = 53, UCL = 88, and LCL = 18. This suggests that these values were established during an interval

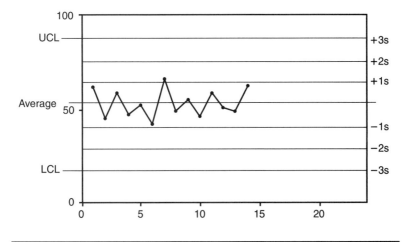

**Figure 4.2** Control chart showing normal variation.

when the process was stable and that the data used in calculating the control limits were not influenced by problems from special causes.

Take care not to lose sight of the fact that this chart, like any control chart, shows values corresponding to a variable or attribute from a control point within our original process. If Figure 4.2 charts the time required to pass a patent application through a Patent Review Committee, then the first data point indicates a review time of 62 hours—or an average review time of 62 hours for the set of patent applications that make up our sample.

Take a closer look at the way the process represented by Figure 4.2 behaves. Notice there is considerable variation around the average, which is 53 hours in this example. Does this suggest something has gone wrong in the process? From this chart alone, it appears the answer is no; after all, not a single data point is above the UCL or below the LCL. Furthermore, it appears that the variations above and below the average value are fairly random. In other words, over the long run there seem to be about as many data points above the average as there are below it—though it is not absolutely essential for adjacent data points to alternately swing above and below the average value for the process to be considered *in control.*

Keep in mind, when speaking of *a problem,* we are referring to a problem that might be induced by a *special cause.* Based on the placement of the control limits, it would appear that the variation shown in Figure 4.2 is due to *common causes* rather than some anomaly. Nevertheless, a client would likely find little consolation in knowing that this stage of the process could require as much as 88 hours, especially if that client had been told that it takes 53 hours on average for a patent application to pass through the Review Committee. If, on the other hand, the UCL had been initially set at 65 hours—implying a higher degree of consistency in this stage of the process—any data point exceeding this value would be attributed to an anomaly. Under these circumstances, the seventh data point in Figure 4.2, with a value of 66 hours, would have been flagged as an out-of-control condition, suggesting the need for corrective action.

This is a good place to point out the folly of trying to overregulate a process that is actually in control but experiences minor variations within the control limits. It can be shown experimentally that making minor process adjustments to offset variation, and thus attempt to limit variation from sample to sample, is an exercise in futility. Again, if we wish to tighten the control limits, fundamental changes in the process itself are called for.

Now that we have seen how a control chart can be used to monitor a process operating under normal conditions, let's examine some of the ways such a chart can be used to highlight an actual, or imminent, problem.

Figure 4.3 shows a situation in which the process seems to be well behaved until a problem suddenly surfaces.

In this case, the chart does not give any warning that a problem is about to occur. In fact, the problem could have been in existence at any time between the point where it became apparent and the time

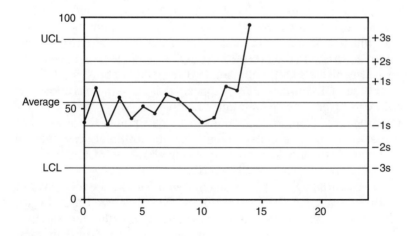

**Figure 4.3** Sudden out-of-control condition.

when the last sample was taken. What might this suggest about our sampling strategy?

Immediate action is typically called for when an out-of-control condition becomes apparent. In other words, we would not wait to collect a series of subsequent samples before taking action, though it is wise to validate the accuracy of our measurements before doing so. If the problem is judged to be sufficiently serious, the entire process may even need to be shut down until the source of the problem can be identified and removed—or at least compensated for by way of an acceptable *work-around solution*. While work-around solutions are sometime unavoidable, be aware that they may shift your process capability, because what we would now call *normal operations* could be something different from what we started with when we constructed the control charts. As a result, new process capability parameters may need to be established. Also be aware that we can expect either control limit to be exceeded about once out of every 700 times due to *common causes* alone, regardless of how undesirable this may be.

Fortunately, SPC is not limited to simply indicating the existence, or nonexistence, of a problem. Perhaps the greater value lies in its ability to map process trends that signal imminent danger. Using the control charts in such a manner, we can direct our focus to *process management* rather than to end-results management. Under the process management approach, we are motivated by the adage: An ounce of prevention is worth a pound of cure. By contrast, end-results management can be only reactive.

Let's now examine several *unnatural patterns* that often signal the early stages of a problem. Be aware these patterns are intended to serve as guidelines rather than as absolute indications of a potential problem. Your own processes may be more, or even less, sensitive to variation than is characterized by these patterns. In any case, it is helpful to examine these patterns to gain an appreciation of the information the control charts are capable of conveying.

Figure 4.4 depicts a pattern suggesting a sudden change in the level of performance of the process.

Such a shift may be caused by changes in some inherent aspect of the process. For instance, in the case involving the Patent Review Committee, a new chairperson may have been installed, resulting in a change in efficiency. In a mechanical system, a system overhaul may cause a sudden shift. Corrective action calls for identifying the cause of the shift and, if necessary, creating a second control chart that will focus attention on the area where the change has occurred. Even if the shift is in a favorable direction, it is still desirable to understand its cause for at least two reasons: to be sure our measurement process has

**30** Chapter Four

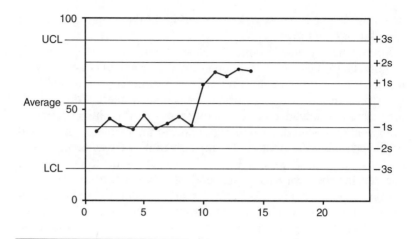

**Figure 4.4** Pattern shift.

not failed and to be able to repeat whatever it is that caused process performance to unexpectedly improve.

Figure 4.5 suggests the existence of some source of *cycling* within the process.

In this case, it appears there is a fairly repeatable and predictable pattern that cycles over a period of time, in contrast to the random variation of highs and lows between adjacent data points that we normally expect. Where they apply, cycling may be caused by such factors as

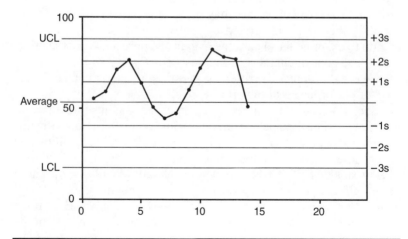

**Figure 4.5** Cycling.

gradual changes in temperature, differences between teams operating in shifts, or high turnover and replacement rates resulting in a cyclical learning curve. Corrective action calls for identifying the source of change and then perhaps identifying other control points to target more precise ways of controlling the process.

Figure 4.6 depicts a gradually shifting trend, perhaps because of some aspect of the process that is sensitive to wear and tear.

Such trends may also occur as the result of an increased demand on the process while faced with a constraint on resources to support the new level of demand. For example, in our patent application process there may be seasonal changes in the number of patent applications received, leading to a backlog in the Patent Review Committee at certain times of the year. If a trending condition is the result of overloaded capacity, we are left with the choice of either living with the condition or determining ways of handling the additional workload, for example, by securing additional temporary support. If we decide to live with the condition until the trend reverses, we may also have to be prepared to deal with the fact that normal variance conditions may occasionally push us beyond one control limit or the other.

Figure 4.7 shows a data plot that appears on the surface to be desirable.

It would appear that the process is under control and that we are experiencing only minor random variations, close to the average value, from sample to sample. Nevertheless, the very fact that the process appears to be working extraordinarily well may be cause for concern. It is possible, for example, that the process samples were not randomly

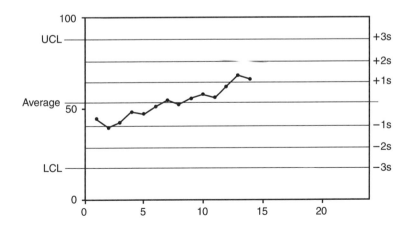

**Figure 4.6**   Trending condition.

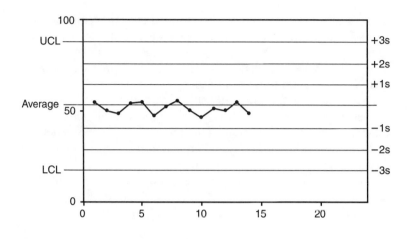

**Figure 4.7** Stratification pattern.

selected, resulting in bias that does not accurately represent the process under all conditions. Consider, for example, how our chart of the Number of Mistakes on the Patent Application Form might be biased if we examine only those patents originating from large companies. If the original process control limits (that is, the UCL and LCL) are truly based on the capability of the process under normal circumstances, over the long run we would expect to see sample values that fall throughout the LCL to UCL range. Sample values that are consistently close to one another, whether they are centered on the average value or elsewhere, should be considered suspect—unless, of course, the process has been improved to make the control limits closer to one another. But, even if the control limits have been tightened, random variation between the new control limits is still to be expected.

The condition just described—where the sample values tend to vary only slightly around a certain level—is sometimes referred to as *stratification*. The use of this term suggests that we may have selected samples that fulfill only certain criteria, rather than samples that are randomly selected from among the entire set of possibilities.

Figure 4.8 depicts a situation in which there is excessive variation in the process. This pattern is typical of what you might expect to see if measurements are blended from two different aspects of the same process or from corresponding elements of two different processes—thus the designation, *mixture pattern*.

An example involving our law firm, Advocate General, will help clarify how such a pattern might occur. In setting up the SPC system

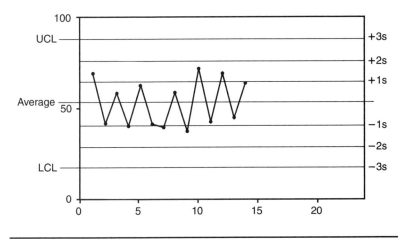

**Figure 4.8** Mixture pattern.

to track and control the time expended in processing patent applications, the SPC Implementation Team found it necessary to separate the data for applications involving hardware products from those involving software products. From historical data, they discovered that software patents took an average of 16 hours longer from start to finish. Had they plotted the data from these two sources on the same control chart, they would likely have obtained a pattern similar to that of Figure 4.8, since the majority of the data points will tend to cluster around the average time of each of these two processes. Furthermore, because the average cycle time for a software patent is 16 hours longer than the average cycle time for a hardware patent, relatively few data points will fall in the vicinity of the center line of the control chart.

It is worth noting at this point that a pattern similar to the one in Figure 4.8 may also occur if the process is consistently *tampered* with in an attempt to compensate for normal variation from one sample to the next. (Recall our earlier remarks about the folly of attempting to microregulate a process that is in control.) If the variation from sample to sample is truly random, large swings from one point to the next may be experienced if adjustments are persistently made solely on the basis of the most recent measurement of process performance. Unless it extends above the UCL or below the LCL, it is typically not wise to tamper with the process based on the value of a single data point.

**Exercise:** *Briefly describe a situation in your own organization in which corrective action appears to be based on the latest results rather than on permanent improvements in the process. (Note: High-profile processes are especially prone to this treatment.)*

_____

_____

_____

_____

**Exercise:** *Based on what you have witnessed, briefly describe your conclusions regarding the effectiveness of this strategy. (For instance, Does it work? Are there any unintended consequences? Are there ways in which the numbers get manipulated?)*

_____

_____

_____

_____

## THE CUMULATIVE SUM CHART

Up to now, we have had little to say about any particular type of control chart. The patterns we have examined could have been plotted on any of the control chart types (which we will later describe) without changing the basic interpretation of the pattern. (Of course, the vertical scales will typically be different from one chart to the next.)

But, unfortunately, none of these charts is very good at indicating a gradual shift in process performance over the long run—especially when there is considerable variation between adjacent data samples. To detect such changes, we take advantage of the CuSum (pronounced Q-sum) chart—which is short for *cumulative sum*. To understand the purpose of the CuSum chart, consider the following analogy:

> An elderly person doesn't simply wake up one day and suddenly discover that her vitality is gone. Yet, over time, significant biological changes occur to diminish vitality. This gradual change in vitality is masked by the fact that each day leading up to the advanced age was more or less different from the previous day. In other words, it is difficult for a person to detect that she is aging while the process is under way—especially considering the fact that physical and emotional vitality naturally vary from day to day.

Let's take a closer look at how the CuSum chart works by reexamining the data points plotted in Figure 4.2 on page 27. We stated that this chart appears to be plotting a process that has normal variation. Nevertheless, let's construct a CuSum chart to see whether the process is perhaps shifting in one direction or the other.

To plot the CuSum chart, we first need to determine how much each of our sample values differs from the target (or average) value. Each point on the CuSum chart represents the cumulative sum of these difference values up to, and including, the most recent difference value. If, for instance, we know that it takes five hours on average to perform a certain task, we will notice a gradually increasing CuSum plot if the task begins to take even a few minutes longer.

Table 4.1 shows the results of the calculations using the data plotted in Figure 4.2.

Notice in Table 4.1 that the first CuSum value is 9, since this is where we first begin calculating and recording our CuSum data points. Looking down the table at the 10th data point, for instance, we see a CuSum value of −5, corresponding to the sum of the current difference value (which is −7) and the previous CuSum value (which is 2).

Figure 4.9 shows a plot of the CuSum values that appear in the right-hand column of Table 4.1.

Notice that the midpoint is zero, as it will always be for a CuSum chart. This is because the CuSum values are referenced to the target (or average) value. Also note that if the process is tracking perfectly, we would expect to see no difference between the sample values and the target value. In Figure 4.9, it would appear, by the slightly downward-trending plot, that this process may be drifting a bit low.

Both of these conditions—stability and trending—are highlighted in the CuSum chart shown in Figure 4.10.

**Table 4.1** Data used in preparing CuSum chart.

| Data point | Sample value | Target value | Difference between sample value and target | CuSum of the differences |
|---|---|---|---|---|
| 1 | 62 | 53 | 62 − 53 = 9 | 9 |
| 2 | 45 | 53 | 45 − 53 = −8 | 1 |
| 3 | 59 | 53 | 59 − 53 = 6 | 7 |
| 4 | 47 | 53 | 47 − 53 = −6 | 1 |
| 5 | 52 | 53 | 52 − 53 = −1 | 0 |
| 6 | 42 | 53 | 42 − 53 = −9 | −9 |
| 7 | 66 | 53 | 66 − 53 = 13 | 4 |
| 8 | 49 | 53 | 49 − 53 = −4 | 0 |
| 9 | 55 | 53 | 55 − 53 = 2 | 2 |
| 10 | 46 | 53 | 46 − 53 = −7 | −5 |
| 11 | 58 | 53 | 58 − 53 = 5 | 0 |
| 12 | 51 | 53 | 51 − 53 = −2 | −2 |
| 13 | 49 | 53 | 49 − 53 = −4 | −6 |
| 14 | 62 | 53 | 62 − 53 = 9 | 3 |

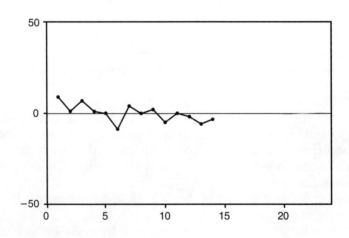

**Figure 4.9** CuSum chart.

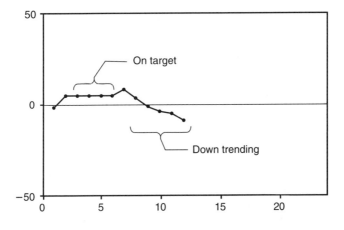

**Figure 4.10**   Stability and trending conditions.

# CHARACTERISTICS OF A NORMAL DISTRIBUTION

Because our discussion has focused on the use of control charts to detect actual or potential problems, it is wise to know when the control charts *do not* indicate a problem. In other words, we would like to know if there is a certain pattern that we might expect to see under normal circumstances.

Recall, when we initially examined Figure 4.2, we made the point that this particular pattern appears to be normal because no data points are above the UCL or below the LCL and there appears to be random variation in our sample values. While these conditions are certainly characteristic of a normal pattern of variation, we need to add still another condition. Specifically, it is important to observe how far the data points fall on either side of the average, even if they are all within the control limits. (This spreading out of the data points is referred to as *dispersion,* and the entire pattern of data points is referred to as the *distribution.*)

If enough samples are taken, it is possible to apply certain principles of statistics to obtain a *best estimate* of where the data points would be expected to fall in relation to the average. (This assumes, of course, that the process is operating under normal circumstances and that we have been *fair,* or unbiased, in selecting our data samples.)

**Table 4.2** Distribution of variable data.

| Range on control chart | Percentage of all data points |
|---|---|
| −1s to +1s | 68.3% |
| −2s to +2s | 95.4% |
| LCL to UCL | 99.7% |

Without getting further into statistical theory, we can make some general comments concerning the position of the data points in relation to the average value under normal operating conditions. Under such conditions, the majority of the data points should lie within the −1s to +1s range of values over the long run. Table 4.2 shows the percentage of data points we would expect to find in each range if the data we are plotting represent variables (as opposed to attributes).

Also note from Table 4.2 that, even under normal circumstances, 0.3 percent of the data points are expected to be outside of the upper and lower control limits. This explains our earlier remark that we can expect one control limit or the other to be exceeded approximately once every 700 times even when no special causes are present.

While the distribution of data points for attributes is more difficult to characterize, the idea is basically the same. In other words, over the long run and under normal circumstances, the majority of the data points will lie within 1s of the average value (that is, between −1s and +1s). Furthermore, there will be at least some, but a fewer number of data points beyond the −1s to +1s range.

In this chapter we have seen how the SPC control charts can be used to visually detect the presence of a problem or impending problem. Once a problem has been identified, we are then left with the task of isolating the root cause. Consequently, when using SPC as a problem-solving tool, it is best to establish control points that monitor *direct* aspects of the process that will put us closest to a trouble spot, rather than *symptom indicators*—at least to the extent this is possible to do. If the process should then go out of control, we will have more definitive data upon which to base our response.

In the next chapter we will discuss how to analyze a process with the ultimate intent of identifying control points within the process. Special attention will be paid to understanding the role of the process team in supporting the process and analyzing its structure.

# 5 | Analyzing the Process Aspects of Your Job

Recall from Figure 2.2 that the first two steps in our model for setting up an SPC subsystem are concerned with determining the process to monitor and then analyzing the process. These steps are often a bit easier to comprehend in a production environment—such as a product manufacturer involved in mass production—than they might in a nonproduction environment, such as a law firm, a marketing firm, an office-cleaning business, or a host of other service or support organizations. But they are equally relevant in any case.

## THE CRITICAL HUMAN FACTOR

Most of us are so conditioned to focus on our immediate job that it is easy to lose sight of the fact that what we do is tied to a larger purpose. Purely and simply, one *integrating purpose* should link every process in the organization: customer satisfaction. Some organizations carry this even further by insisting on nothing less than *total* customer satisfaction. Furthermore, when the focus is truly on the customer, individual employees will be more likely to adopt a process point of view rather than seeing their jobs in isolation.

Battles involving turf, job specialization, departmental duties, and a *not-invented-here* mentality have little place in an organization where the emphasis is on serving the customer's needs. Nevertheless, these factors persist in many organizations—especially service and support organizations. As a result, we may have to radically alter our point of view to begin even thinking of our jobs as part of a process that is designed to translate customer needs into practical solutions. It should go without saying that interdepartmental barriers need to come down if we are serious about a *process management* style of doing business, because most every process cuts across departmental lines to some degree.

Why should this concern you if are an individual contributor, rather than a manager? For the simple reason that everyone involved in the process must believe in the customer satisfaction way of doing business and that they have a direct contribution to the process and its success. It should also be clear that in this highly competitive age, any business that is not customer focused is, sooner or later, destined to fail. As a result, such issues as turf-preservation mean little in the long run.

With the current emphasis on permeating quality throughout every aspect of the organization, senior managers are rapidly gaining an appreciation for the fact that the responsibility for defining the various processes, and the roles of teams and individuals, will necessarily involve everyone, especially those who support the process day to day. Cross-functional task forces are often formed to plan and analyze the various processes throughout the organization. Quite often such a task force will work under the guidance of an individual who has been identified as the "process owner." The process owner is often the individual—typically a manager—who has the most resources committed to a particular process and who has the largest responsibility for the quality of the outcome.

So what might you, if you are a process team member, be expected to contribute beyond your traditional contribution as a specialist or one who fulfills a support role? The answer will naturally depend on a number of factors, such as the size of the organization and the nature of the products or services provided to the customer. The following items are some of the process management tasks that process team members have been called on to perform:

- Identifying the components of the overall process (that is, subprocesses, inputs, outputs, key work activities, and task responsibilities)

- Identifying how the process flows—in other words, how work, information, approvals, and so forth flow through the process

- Identifying critical interface points between the various components of the process (for example, places where things can "slip through the cracks" or become delayed because of responsibility voids)

- Classifying which aspects of the process do, and do not, add value to the process (value-added elements include direct features of the product or service that the customer is most interested in; non-value-added factors include delays, storage, and transportation)

- Identifying cost factors, and areas of improvement, related to accomplishing all of the performance factors that are important to the customer (that is, the "cost of conformance") and nonconformance, such as waste, inefficiency, rework, and unnecessary delays.
- Identifying control points within the process that can be expected to have the greatest impact on process performance (that is, weak links)
- Planning the SPC subsystem (for instance, establishing baseline performance and determining which factors to monitor) and later collecting data to prepare the control charts
- Solving problems that are due to special causes—perhaps from interpreting abnormal conditions by using the control charts
- Identifying ways in which the process can be made more efficient, improving its ability to translate customer needs into solutions (that is, continuous improvement by eliminating common causes of problems)

At first glance, it may appear that these tasks will add to your overall workload. In reality, the opposite often occurs. For instance, if work orders do not have to be handed back and forth between departments, not only will the process cycle time be reduced, thus pleasing the customer, but everyone involved also will realize a reduction in wasted time because the process is now running more smoothly than before. Certainly, everyone's frustration level will be greatly reduced—a significant benefit unto itself.

Returning to the example of the Advocate General law firm described in Chapter 3, let's review the contribution of the process team members toward planning and analyzing the patent application process. In this case, the process team members may have been involved in some or all of the following:

- Participating in training to become familiar with the purpose for moving toward a process management way of doing business, to understand their contribution toward planning and supporting the SPC subsystem, and to acquire skills in problem solving by using SPC and related tools
- Analyzing the process and constructing the process flow diagram, as shown in Figure 3.1, and perhaps other diagrams, such as an *information flow diagram*

- Identifying an array of possible control points within this process and then selecting the three critical control points that the team identified in Chapter 3
- Agreeing on the most appropriate factor (or set of factors) to be monitored, corresponding to each of these three control points, including the most appropriate SPC tool for charting each factor
- Determining the best way in which to monitor and report process performance—that is, how, how often, by whom, and so on
- Establishing the average values and the control limits (UCL and LCL) by examining historical data

## MAPPING THE PROCESS

Perhaps the most useful tool in performing a process analysis is the process flowchart. Figure 3.1 is only one of several versions of such a chart. Process flowcharts have the potential of fulfilling several important functions:

- They provide a visual tool for depicting the process, making it relatively easy for team members to detect any false assumptions regarding the way in which the process truly functions (for example, missing subprocesses or false assumptions regarding interface points).
- They serve as a convenient team communications tool for detecting critical control points and/or bottlenecks within the process.
- They make it clear how and where the process cuts across functional lines, suggesting areas for improvement and cooperation.
- They can be made to distinguish between the value-added portions of the process and the non-value-added portions.

We should note at this point that classifying certain parts of the process as being "non-value-adding elements" does not suggest that they are inessential. For example, transporting a product or service to a client may be essential to the process, but, from the customer's point of view, it does not directly add any features or functions to the product or service itself. While it is typically not possible to eliminate the non-value-added elements, such elements are especially important to consider when process improvement is sought.

Figure 5.1 illustrates a special adaptation of the process flowchart. As you can see, this particular format makes it clear who (that is, individual, team, or department) is responsible for each of the various subprocesses. As such, it clearly identifies the interface points that are often the source of problems in and of themselves. A similar chart could be constructed to show the flow of information from point to point within the process.

As we have stressed in this chapter, the human dimension of SPC is vitally important to the success of any SPC subsystem. People need to believe that SPC will truly help them better perform their jobs and that they are an integral part of the process itself. There is no better way of doing this than by getting every member of the team who routinely supports the process involved in the planning and analysis. In most cases these are the same people who best understand the process and can provide critical insight into possible process improvements. As an added benefit, organizations often find this effort to be fruitful in identifying and instituting certain rapid improvements—such as eliminating unnecessary delays and bottlenecks—even before proceeding further with the implementation of SPC.

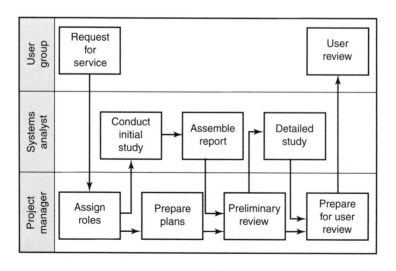

**Figure 5.1** Process flowchart example.

# 6 Deciding Where to Focus Your Effort

Even with the most basic process, we may be faced with numerous choices when selecting the particular aspects we wish to measure and monitor as an indication of process performance. Furthermore, when performing a process analysis, we are likely to uncover even more possible data collection points as the process is broken down into greater levels of detail. Added to this is the fact that we may be able to identify several types of measurements for every control point.

It is easy to see that the task of deciding where to focus our attention when making measurements could become monumental. Our interest in this chapter is in briefly examining several tools and techniques that can make this task more manageable. Specifically, we will examine the following tools and techniques in the context of identifying and selecting process performance indicators:

- Brainstorming
- Fishbone diagram
- Pareto analysis

## BRAINSTORMING

Several guidelines can be followed to make the process of *brainstorming*, which has been around for a number of years, more effective than simply getting a group of people together and randomly tossing out ideas. These are:

1. Enlist the help of a facilitator, who will ensure that everyone is given equal opportunity to provide input and prevent censoring of ideas; record participants' responses as ideas come forth; and try to maintain some degree of

order during the session, while insisting that all dialogue be directed to the entire group.

2. Ensure that the intent of the brainstorming session is clear from the outset and that the focus of the session is sufficiently narrow so that the issue you are interested in will indeed be addressed (a single issue should be addressed in each session).

3. Insist that no one pass judgment on the merits of a particular idea while the ideas are being generated, either by making direct comments or by projecting negative body language.

In terms of identifying measurement points within a process, brainstorming may be used to secure a wide range of untested responses to questions such as the following:

- What characteristics of our product and/or service do our customers consider to be most important?
- What positive (or negative) factors distinguish us from the competition?
- What are some possible ways of measuring _____?
- Where do problems surface within the process?

Notice that we have attempted to frame these questions without injecting bias. For instance, it is better to ask, "Where do problems surface within the process?" than to ask, "Where do problems *most often* surface within the process?" Even though we do ultimately wish to know where problems most often surface, adding this qualifier during the brainstorming session could suppress certain ideas that deserve consideration.

## FISHBONE DIAGRAM

With the fishbone diagram, we have a tool for generating a wide range of ideas regarding possible causes of an actual, or potential, problem. For this reason, the fishbone diagram is sometimes referred to as a cause-and-effect diagram. Others refer to it as an Ishikawa diagram, in honor of its originator, Kaoru Ishikawa. Even though ideas are openly elicited, the fishbone diagram offers a more structured approach to examining a particular problem than that typically

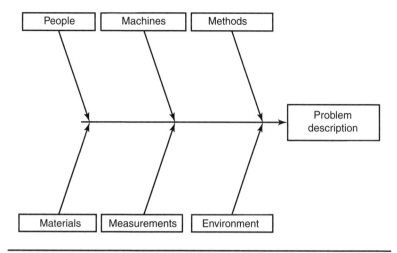

**Figure 6.1** Fishbone diagram.

accomplished during a brainstorming session. (On the other hand, brainstorming is not limited to generating ideas regarding possible causes of a problem.)

As you can see from examining Figure 6.1, the fishbone diagram gets its name from its appearance. Refer to this figure as you study the following recommended procedure for using this particular tool:

1. Agree on the problem you wish to examine in terms of its possible root causes; then write a brief problem description in the box (or fish head) on the right side of the diagram.

2. Create branches off the "backbone" and add headings to these branches to describe an array of related problem causes (start with the people, machines, methods, materials, measurements, and environment categories if you don't have a more specific set of categories in mind).

3. As ideas are elicited regarding possible problem causes, add sub-branches under the major branches to depict which category each idea relates to.

4. Branches may be added beneath other branches, to the degree necessary, as the problem is examined in greater detail—or a second fishbone diagram may be

constructed to examine one particular problem cause in depth by making that problem the fish head for the new diagram.

5. Once you feel you have explored all of the possible problem causes, the most significant causes can be circled (or otherwise flagged) on the diagram.

While the fishbone diagram can be used to uncover possible causes once a problem has occurred, our interest is in knowing how to use this tool to set up our SPC subsystem by narrowing the focus of our effort. There is little difference between the two applications—except for the fact that one is reactive and the other is proactive. In other words, in setting up an SPC subsystem, we might use the fishbone diagram to elicit answers to the following questions:

- What various causes could give rise to a particular type of problem?
- How are these causes related to other causes?
- Of these causes, which ones are most likely to lead to the problem condition that is being considered?

Once we have answered these questions, we would try to determine the following:

- Where in our process might we expect the most likely problem causes to become manifest?
- What direct and/or indirect measurements should be made to detect the presence of these causes?

One variation of the fishbone diagram is shown in Figure 6.2. This particular variation is sometimes called the process classification format. As you can see, this format is essentially the same as the process flowchart but with branches added during the analysis to show where problems are most likely to occur. Notice too that branches can be added to the subprocesses as well as the *interfaces* between the subprocesses—those important interchanges where things fall through the cracks, often because of miscommunication or faulty assumptions regarding who has responsibility for certain actions.

The example in Figure 6.2 shows a certain process involved in filling a customer order. Each area identified on either side of the process points to possible problem areas—suggesting control points that should be monitored and measured using an SPC subsystem.

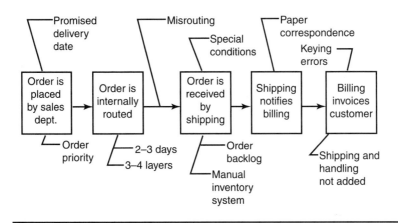

**Figure 6.2** Process classification format.

Notice also that we have flagged the interface between the two subprocesses designated as "order is internally routed" and "order is received by shipping" as a possible area of concern, denoted in this example as "misrouting."

## PARETO ANALYSIS

Named in honor of economist Vilfredo Pareto (pronounced p ə-rā'tō), Pareto analysis is helpful in determining where to focus our effort when planning an SPC subsystem. This is especially relevant to situations in which numerous control points can potentially be monitored and we are constrained by available resources in our ability to do so. Furthermore, if this tool is used as a means of ranking problem causes according to their frequency of occurrence—as it often is—the information obtained may later help identify the underlying causes of problems that appear as anomalies on the control chart.

Pareto analysis is based on the familiar *80/20 rule*—also called the *Pareto principle*—which states that 80 percent of the problems that occur in a certain process or product can be attributed to as little as 20 percent of the possible causes. Though in reality the ratio of 80 to 20 may not be exact, it's almost always the case that relatively few causes are responsible for a disproportionately large number of the negative consequences. Also, because statistical process control is concerned with avoiding problems *before* they occur—or possibly lead to even greater problems—we are especially interested in monitoring high-incidence causes that provide early indication of an impending problem.

Figure 6.3 shows a Pareto chart that was prepared by our law firm's SPC Implementation Team. In preparing this chart, the most recent historical data were collected to answer the question: "What are the sources of errors in patent applications that are submitted to the patent office?" (Details on how this chart was constructed are provided in Appendix A.) In this example, *errors* are the problem and *sources of errors* are the causes.

Notice from Figure 6.3 that the causal factors are arranged in order from high to low, according to the frequency in which they occurred over the two years. Also notice that each point on the curved line corresponds to the sum of the errors originating from the source of errors up to that point, expressed as a percentage of the total. This example clearly demonstrates the Pareto principle in action, showing that a small number of causes are responsible for most of the problems and that the relative frequency of errors tends to diminish rapidly

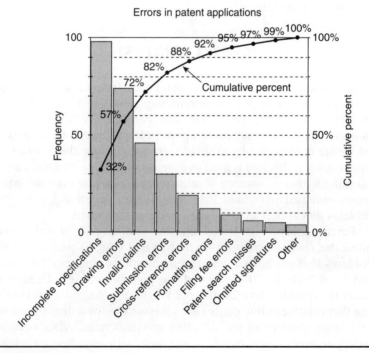

**Figure 6.3** Pareto analysis example.

in moving from the left to the right side of the chart. In this case the first three factors—incomplete specifications, drawing errors, and invalid claims—account for 72 percent of the total number of errors, assuming these errors are independent of one another.

Before moving on, it is worth noting that Pareto analysis can also be used to determine areas where positive results are occurring, following the reverse but equally valid notion that a relatively small number of factors tends to have a disproportionately large influence on desirable outcomes. In other words, if our process is doing something particularly well, we can use Pareto analysis to identify the primary factors that are contributing to our strengths and then seek to have them reinforced. While this is especially important for process improvement, it may also help identify certain positive factors we may wish to track with the aid of control charts—for instance, a customer satisfaction rating in a certain performance area that is critical to our company's competitiveness.

The previous example highlights the often overlooked fact that process control has two faces: one directed toward avoiding *undesirable* outcomes, the other toward sustaining *desirable* outcomes. When considering process control opportunities, it is important to recognize that one is not necessarily the inverse of the other—for instance, the lack of an unhappy customer does not necessitate the presence of a delighted customer. Paying attention to one and not the other is akin to an artist painting a foreground and ignoring the background.

# 7 Barebones Statistics

We have been able to come this far in our examination of statistical process control without needing to understand statistical theory, other than perhaps intuitively. While we intend to continue along this path as much as possible, we do need to clear up a few—and we really do mean a few—statistical concepts in order to describe the specific types of control charts that we will examine in the following chapter.

We also need to clarify something called "standard deviation" so we can be more specific than we have been so far in determining the upper and lower control limits. Determining the standard deviation involves a mathematical operation that most of us seldom use in our day-to-day lives: taking the square root. If your pocket calculator has a square root ( $\sqrt{\phantom{x}}$ ) key, as even the most inexpensive pocket calculators do, determining the square root is simply a matter of entering the numbers correctly and then pressing the $\sqrt{\phantom{x}}$ key.

If you are comfortable with basic statistics, you may wish to skip or skim the remainder of this chapter.

## AVERAGES AND MEANS

When we introduced the control chart format in Figure 2.3, we indicated the need to identify three values in order to construct the chart: upper control level (UCL), lower control level (LCL), and average. Let's say a few words about the meaning of the term *average value* to ensure that we are speaking the same language when we refer to this term in the future.

The average value is a simple calculation that requires adding a set of numbers and then dividing this sum by the number of items that we added together. If, for example, you have three chocolate squares weighing 5 ounces apiece and six chocolate squares weighing 2 ounces apiece, the average value would be determined as follows:

1. Total the weight of all chocolate squares: 27 ounces

    Because $(5 + 5 + 5) + (2 + 2 + 2 + 2 + 2 + 2) = 27$

2. Number of chocolate squares: 9

3. Average value: $27 \div 9 = 3$ ounces per square

What does this mean? It simply means if we had some way of dividing the nine squares of chocolate into equal sizes, then each piece would weigh exactly the same: 3 ounces.

With something like chocolate, this would be fairly easy to do, because we could simply melt and combine all of the chocolate and then cut the batch into nine evenly divided pieces once it cools and hardens. But, with something like tires, this is not so easy to do; nonetheless, the meaning of the term *average value* is the same in either case.

There is still another way of thinking about the *average value* that is especially relevant to SPC. Let's say, for instance, that you were to drop the original nine squares of chocolate into a brown paper bag. If you were to tell someone else nothing more than the fact that the bag contains nine chocolate squares and that the entire bag weighs 27 ounces, what would be their best guess of how much each chocolate square weighs? The answer, of course, is the average value: 3 ounces. In other words, this is what they would *expect* the chocolate squares to weigh if they knew nothing else about the individual squares, even though we are privy to the fact that not a single chocolate square in the bag actually weighs 3 ounces.

Why, you may ask, is this point of view especially relevant to SPC? Perhaps an example can help clarify the situation.

Again, consider the hypothetical law firm introduced in Chapter 3 (Advocate General). Let's say that the firm checked its historical records and found that the Review Committee required an average of 36 hours to complete a patent review. Knowing this, what is the best estimate the firm could give a potential client about how long the Review Committee would need to review the client's application? The best estimate is the average value, 36 hours, even though the actual time may fall within a wide range of values on either side of 36 hours.

The symbol $\overline{X}$ is used in SPC as a shorthand way of referring to the average value. We call this symbol X-bar. In other words: $\overline{X}$ = X-bar = average value.

Before leaving the discussion on average value, we should introduce the term *mean value*. When we speak of the mean value, we are referring to exactly the same thing as the average value, at least as we defined *average value* above. Technically, *mean value* is more accurate in uniquely describing what we have, up to now, referred to as the *average value*. However, for the most part, *mean value* and *average value* are used interchangeably.

## RANGE VALUES

In SPC, when we speak of the *range* of a set of numbers, we are referring to the difference between the largest number and the smallest number in that set. (Notice that this definition of *range* implies a single numerical value, rather than a from/to set of values.) Consider, for example, the earlier situation involving the chocolate squares. Because we know that the largest square weighs 5 ounces (even though there are several that weigh this amount) and the smallest square weighs 2 ounces, we are able to determine that the range is 3 ounces (5 ounces minus 2 ounces).

Knowing the range value of a set of numbers, as well as the average value, gives us additional information to work with. Let's refer again to the case involving the law firm to see why.

Suppose that in addition to telling the client that the Review Committee takes an average of 36 hours to complete a patent review, we were to relate the fact that such reviews have taken as long as 48 hours and as little as 8 hours. With this additional information, the client has a better grasp of the situation. In this case the range is 40 hours (48 hours minus eight hours).

Also note that the difference between the maximum value (48 hours) and the average value (36 hours) is 12 hours, which is less than the difference between the average value (36 hours) and the minimum value (8 hours), which is 28 hours. In other words, the range is not symmetrical on each side of average value. Undoubtedly the client would find this information helpful as well.

In SPC, we use the shorthand symbol R as an abbreviation for the range value. Note that R = 0 any time the set of numbers under consideration all have the same value. Also remember to keep straight the distinction between the word *range* when referring to a single value—as it often does in SPC—and its everyday use in referring to a pair of from/to values, such as the miles-per-gallon your family car gets in town versus the open road.

## STANDARD DEVIATION

Similar to the range value, the standard deviation is simply another way of calculating the degree of variation (or spread) that exists within a set of numbers. The steps involved in calculating the standard deviation are as follows:

1. Determine the average value (X-bar) of the set of numbers.
2. Find the difference between each individual number and the average value.
3. Square each of the difference values found in step 2.
4. Add the squared difference values found in step 3.
5. Divide the total found in step 4 by the number of original items.
6. Take the square root of the value found in step 5.

Let's apply these steps to some numbers to help clarify the meaning of standard deviation. Say, for example, that we would like to determine the standard deviation given the set of numbers 6, 2, 8, 12, 10, and 16. Following the six steps above we get:

1. X-bar = 9

    Because $(6 + 2 + 8 + 12 + 10 + 16) \div 6 = 9$

2. The difference values are:

    $6 - 9 = -3$
    $2 - 9 = -7$
    $8 - 9 = -1$
    $12 - 9 = 3$
    $10 - 9 = 1$
    $16 - 9 = 7$

3. Squaring each of the difference values we get:

    $(-3)^2 = 9$
    $(-7)^2 = 49$
    $(-1)^2 = 1$
    $(3)^2 = 9$
    $(1)^2 = 1$
    $(7)^2 = 49$

4. Adding these squared difference values we get:

    $9 + 49 + 1 + 9 + 1 + 49 = 118$

5. Dividing this sum by the number of items we get:

    $118 \div 6 = 19.67$

6. Taking the square root of this value we get:

    $\sqrt{19.67} = 4.43$

Keep in mind that the standard deviation is simply a term applied to a calculated value that tells us something about the degree of variation that exists within a certain set of numbers. As we see from step 2, the reference point for standard deviation is the average value. Also, the larger the standard deviation value, the more variation we would expect to find within the group of numbers.

If, for instance, we calculate the standard deviation for the set of numbers 4, 2, 8, 12, 12, and 16, we obtain a value of 4.86. Several interesting points are worth noting when comparing this set of numbers with the first set of numbers. Consider the following:

| Set of numbers | Average | Standard deviation | Range |
| --- | --- | --- | --- |
| 2, 6, 8, 10, 12, 16 | 9 | 4.43 | 16 − 2 = 14 |
| 2, 4, 8, 12, 12, 16 | 9 | 4.86 | 16 − 2 = 14 |

Notice that except for two values, the second set of numbers is identical to the first set. In fact, the second set of numbers was created by decreasing one of the numbers in the first set by 2 while increasing one of the other numbers by this same amount. As a result, the average value remained the same. Also, the range did not change, because the maximum and minimum values did not change. What did change was the standard deviation. In other words, the standard deviation was able to detect the fact that there is more variation *within* the second set of numbers than there is within the first set of numbers. (Again, when we say *more variation* we are referring to the variation in relation to the average value, or mean, which is 9 in both cases.)

Thus, as this example shows, while the range and the standard deviation both indicate how much *dispersion* there is in a set of numbers, the standard deviation is more sensitive to variations *within* the set of numbers.

Also be aware that the letter s is used as a shorthand designation for standard deviation. As we saw in Figure 4.1, it is common practice in SPC to mark off distances from the average value in increments of s (that is, +1s, +2s, UCL, −1s, −2s, and LCL). Again, the upper control limit (UCL) is 3 standard deviations above the average value, while the lower control limit (LCL) is 3 standard deviations below the average value. Note, too, that the lowercase Greek letter sigma ($\sigma$) is often used as a shorthand designation for *standard deviation*—which is where the *sigma* in the Six Sigma quality philosophy comes from.

If we were to create a control chart using the first set of numbers above, we would mark off the following values on the chart:

1. Average = X-bar = 9
2. X-bar + 1s = 9 + 4.43 = 13.43
3. X-bar + 2s = 9 + 8.86 = 17.86
4. UCL = 9 + 13.29 = 22.29
5. X-bar − 1s = 9 − 4.43 = 4.57
6. X-bar − 2s = 9 − 8.86 = 0.14
7. LCL = 9 − 13.29 = −4.29

(The LCL is typically set to zero if the computed value is a negative number, as it is in this case.)

One final point should be made before moving on. At the beginning of this section we walked through a sequence of steps—or algorithm—for calculating the standard deviation. This is fine for instructional purposes but it can be a daunting task in practice. Fortunately the standard deviation of a set of numbers can be easily calculated with an inexpensive pocket calculator (look for keys labeled $\sigma$ and s).

## POPULATIONS AND SAMPLES

The term *population* refers to the entire set of values, or items, that constitute a certain set of factors we happen to be interested in. For instance, the population of males between the ages of 16 and 18 years of age would include every young man in the 16-to-18 age group.

Using the law firm example in Chapter 3, if we were to speak of the *population of time delays* involved in processing patent applications, we would be referring to the time delay associated with every possible patent application. While it is at least conceivable that we could determine for sure whatever it is we are interested in knowing about the population of males between the ages of 16 and 18 (such as their average height), it is impossible to make an absolute statement about the population of patent application delays, because this particular population will include patents to be processed in the future as well as those that have already been processed.

The latter situation is the more common of the two; in other words, we have to live with incomplete information about the population in most cases. As a result, we typically rely on *samples* taken from the population in order to estimate the characteristics of the population itself.

Even when complete information is available, the sheer volume of information represented by the population may be more than we can handle. For instance, though it may be within the realm of possibility to ask everyone in the United States how they will vote in a presidential election, the sheer magnitude of this task suggests the need for querying only a representative sample of the population of voters. By a *representative* sample, we mean a sample in which the individuals are randomly selected and one that is sufficiently large enough that we are assured, within some allowable margin of error, that the sample characteristics faithfully depict the population characteristics.

Keep in mind that a population can consist of any classification of items that share a common characteristic we are interested in counting or measuring. For instance, it is possible to conceive of a population consisting of values that are calculated from an infinite number of samples that are randomly drawn from some other population. In the population of 16- to 18-year-old males, for example, we can imagine another population of values corresponding to the average height, as computed by randomly selecting 10 young men at a time.

In fact, a population consisting entirely of values that are calculated by averaging samples that are drawn from another population (such as a population of individual measurements) is of particular interest to SPC. Such a population—one consisting entirely of sample averages—will exhibit the characteristics of the well-known *bell-shaped curve*. This bell-shaped curve, also referred to as a *normal distribution*, makes it possible to perform certain calculations and estimations that would be meaningless otherwise. For example, the distribution of data points in Table 4.2 is based on a normal distribution.

## NORMAL DISTRIBUTION

One of the most important features of the bell-shaped curve, or normal distribution, is the fact that we can use a special set of tables to determine the percentage of values that fall between any two points on the scale of values described by the curve. If, for instance, we are able to determine that the population consisting of the height of males between the ages of 16 and 18 years of age is defined by a normal distribution, it is possible to determine how many males in this age group fall within the range of 5 feet, 9 inches and 6 feet, 2 inches. To make this determination we need to know three facts:

1. That the distribution of values is in fact normal (bell-shaped)
2. The average value
3. The standard deviation

We know, for instance, that 68.3 percent of the values defined by a normal distribution will fall within 1 standard deviation on each side of the average value. For this same distribution, we also know that 95.4 percent of the values fall within 2 standard deviations of the average and that 99.7 percent of the values will fall within 3 standard deviations of the average.

In other words, under normal circumstances—that is, under normal operating conditions—the upper and lower control limits on an SPC chart will cover 99.7 percent of the possible values for a particular factor we are monitoring, but only if it is possible to assume that the values are selected from a normal distribution. If our control chart is based on sample values, as opposed to population values, this is typically a valid assumption.

If, for example, we had some way of knowing that the average height of males between the ages of 16 and 18 years of age is 5 feet, 7 inches and that the standard deviation of this particular population is 3 inches, it is possible to use this information, and a set of tables, to determine that 24 percent of the population of 16- to 18-year-old males will fall within the range of 5 feet, 9 inches to 6 feet, 2 inches in height. These tables—which can be found in most statistics textbooks—contain values that have been calculated from the equation that defines the *shape* of the normal distribution. Therefore, using this method to determine the percentage of a population that lies between any two values is only valid if the population has a normal distribution—as is often the case in so-called *natural systems*.

Figure 7.1 demonstrates this in graphical form. As you can see, the average height (5 feet, 7 inches) is shown at the peak of the bell-shaped curve. The area under the curve between 5 feet, 9 inches and 6 feet, 2 inches represents 24 percent of the area under the entire curve, which is equivalent to saying that 24 percent of the entire population falls within this range. The curve also indicates that very few 16- to 18-year-old males are taller than 6 feet, 2 inches, because the area under the curve beyond 6 feet, 2 inches is small in comparison to the total area.

In general, a plot such as Figure 7.1 is known as a *probability distribution*. For each point on the horizontal axis (which is a particular height in this case), it shows how often (or, more accurately, what percentage of the total) a particular value is represented. With a normal distribution, the average value always represents a higher percentage of the total than any other value.

## STATISTICS AND PARAMETERS

Now that we have described the difference between a population and a sample, we need to distinguish between two additional terms: statistics and parameters. A *statistic* is some quantifiable characteristic of a sample, and a *parameter* is some quantifiable characteristic of a population. For instance, the standard deviation of a sample is a statistic denoted by s and the standard deviation of a population is a parameter denoted by σ. Because we often do not know the value of a particular parameter, we typically have to rely on statistics, such as the average of a set of range values, to approximate the parameter of interest.

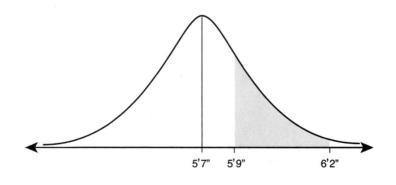

**Figure 7.1** Normal distribution example.

Again, if we were to measure the height of every male in the 16- to 18-year age group, we could derive a parameter representing the average height of this particular population. If our average value is based on a sample, rather than the entire population, then we are dealing with a statistic rather than a parameter. In other words, a statistic relates to a sample in the same way that a parameter relates to a population. In the realm of SPC, we mostly deal with samples, and thus statistics. As with a parameter, note that it is possible to create a population corresponding to a particular statistic, such as a population consisting of the average values of a set of samples.

We can now summarize the following points regarding the use of statistics and parameters to construct our control charts:

1. If there is reason to believe that whatever it is we wish to measure will produce a set of values that display a normal distribution, we can establish *statistically valid* control limits based on the standard deviation ($\sigma$). In most cases, a reasonably accurate estimate of $\sigma$ can be obtained by calculating the standard deviation of at least 30 randomly selected measurements that are taken when the process is (or was) operating normally. Furthermore, spacing the control limits at $3\sigma$ on each side of the centerline is an accepted convention for SPC charts that strikes a balance between putting us in a position of overreacting to normal process variation and that of potentially missing a problem condition altogether.

2. If we cannot be assured that the entire set of data points (that is, the population of values) would be represented by a normal distribution, we can still generate statistically valid control limits by working with sample averages. In this case, the values plotted on the control chart are the sample averages, as opposed to individual data values. Because we are plotting sample averages, our UCL and LCL values will be offset from the average value by an amount equal to three times the standard deviation of the individual values divided by the square root of the sample size. In other words, in the case of sample averages, the larger the sample size, the closer the control limits will be to the centerline (or average value).

In this chapter we have attempted to stay true to our goal of making SPC accessible to our primary audience—right-brain thinkers—without getting overly bogged down in statistical theory. Nevertheless, we hope that what you have learned will pique your interest in learning more about the statistical foundations of SPC. Listed in the References at the end of this book are several "made-easy" books that you may find helpful in furthering your knowledge.

# 8 Planning Your Charting and Measurement System

In Chapter 4 we examined control chart patterns that we might expect to see under both normal and abnormal conditions. At that time, we did not distinguish between the various types of control charts. We simply made the point that the interpretation of a particular pattern is independent of the type of control chart. A trending condition, as in Figure 4.6, would appear much the same whether we were examining what we will later define as a P-chart or a C-chart.

Our focus in the current chapter is on examining six types of control charts. These are the X-bar chart, the R-chart, the NP-chart, the P-chart, the C-chart, and the U-chart. (Note, these charts are sometimes designated by lowercase letters, such as p-chart rather than P-chart, because they plot statistics rather than parameters.)

First, we need to review the fact that we might expect to see two classifications of data: variables and attributes. The distinction between variables and attributes, as described in Chapter 4, is worth noting again and slightly expanding on:

*Variables* represent measurable data that can assume virtually any numerical value, typically within a specified range. The term *measurable* suggests that some sort of instrument—such as a clock for measuring time—will most likely be involved in collecting such data.

*Attributes* represent nonmeasurable data that can be tested only in terms of the presence, or absence, of a certain quality (such as defects versus no defects, pass versus fail, light versus dark, glossy versus dull). Rather than making measurements, with attributes we are likely to count how often (or what percentage of the time) the attribute in question is present. Note that human judgment, complete with a certain amount of subjectivity, may enter the scene with attribute data.

Variable data are most often (though not exclusively) associated with a production environment—such as a product manufacturer—where extensive measurements are taken and specialized instruments are available for doing so.

Attribute data, on the other hand, are more frequently (but not exclusively) associated with a nonproduction environment, such as a service or support organization.

Variable data are often preferred over attribute data, because variable data can provide a means of measuring the degree of variation in the process (though some would argue that any deviation in the process is undesirable). Bear in mind, anytime an instrument as simple as a clock or a ruler is involved, we may have the basis for collecting variable data rather than attribute data, regardless of the work environment or the type of organization.

## THE X-BAR AND R-CHARTS

The X-bar chart and the R-chart are discussed here in the same context because they are often used together to detect an actual, or impending, problem condition. Nevertheless, they are two distinct charts that tell us something quite different about the process under scrutiny.

Both charts are used to monitor variable data—in other words, data obtained via measurement. When a sample is taken, we compute the average value of the sample and also determine the range corresponding to the data items within that sample. As shown in Figure 8.1, the sample averages are then plotted on an X-bar chart while the range values are plotted on an R-chart. When interpreting these two charts, keep in mind that the values plotted on the X-bar chart are based on samples and therefore provide an estimate of where the average value of the process is *positioned* at any given time. Likewise, the values plotted on the R-chart provide an estimate of the degree of *spread* of the process from which the samples were taken.

Recall the fact that control limits (UCL and LCL) and the average value for each control chart are initially established by observing the process under normal conditions—perhaps by carefully controlling the process while collecting baseline data. Consequently, the baseline average for an X-bar chart is computed by taking the average value of a set of sample averages, or the average of a set of averages, so to speak. As a result, the baseline average for an X-bar chart is sometimes referred to as the *grand average* or the *grand mean* and

# Planning Your Charting and Measurement System 65

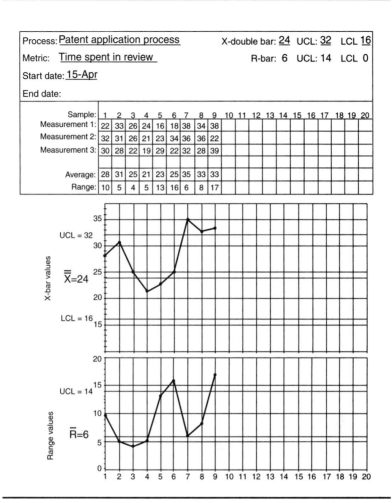

**Figure 8.1** X-bar and R-chart example.

it is depicted by an X double-bar ($\bar{\bar{X}}$) symbol. The average for the R-chart is simply the average of the sample range values used in constructing the chart. As you can see, this is referred to as R-bar ($\bar{R}$).

Recall from the previous chapter that we described how to establish the upper and lower control limits (UCL and LCL). This procedure applies to both the X-bar chart and the R-chart. Be aware that it is also possible—and common practice—to establish the control limits by using the R-bar value and a special set of tables that allow for estimating the standard deviation on the basis of the sample size and the R-bar value. This method of establishing the control limits is typically easier than actually calculating the standard deviation, but it is less efficient if large samples are involved. (Refer to the Amsden book in the References—or virtually any standard textbook on SPC—if you wish to learn more about this method.)

Notice, in Figure 8.1 on the previous page, how we computed the average and range for each sample and then plotted these data points on our X-bar and R-charts. In studying this chart, note that one of the following four conditions could exist when we link the X-bar chart with the R-chart:

1. Both charts could indicate an in-control condition for a given sample (samples 1 through 5).

2. The R-bar chart could indicate an out-of-control condition while the X-bar chart is in control (sample 6).

3. The X-bar chart could indicate an out-of-control condition while the R-bar chart is in control (samples 7 and 8).

4. Both charts could indicate an out-of-control condition (sample 9).

The second condition could occur if there is a considerable difference in the individual values within a certain sample, even though the corresponding average value of the sample is acceptable. Consider, for example, the two sets of sample values from Figure 8.1 that are shown in Table 8.1.

Notice that even though these two samples both have an average of 25, the *range* of the data values in the second sample is four times that of the first sample. This suggests that if we were to track our process by simply using the X-bar chart, we might not detect wide differences between the data points in a given sample, because such differences tend to cancel each other out when calculating the average value. In other words, we may be deceived into thinking that the process is in control when, in fact, the data values in a particular sample might range from very high to very low. Seeing that the time-

**Table 8.1** Comparison of sample statistics.

| Sample | Data Values | Average | Range |
|---|---|---|---|
| 3 | 26, 26, 22 | 25 | 26 − 22 = 4 |
| 6 | 18, 34, 22 | 25 | 34 − 18 = 16 |

spread of the measurements in sample 6 is much larger than expected, we would want to investigate why. In doing so, we may determine that something has changed in the patent application process that allows some applications to be expedited, while others experience inordinate delays.

Now that we have seen how the X-bar and R-charts work in conjunction, let's summarize what the four conditions might indicate:

1. If both charts are in control, the process would appear to be stable and functioning normally.

2. The R-chart being out of control while the X-bar chart is in control would indicate considerable variance within each sample. This could be caused by measurement error, or we may in fact have a very erratic process.

3. The X-bar chart being out of control while the R-chart is in control would indicate that there is little variance within each sample, suggesting that our process truly has drifted out of control.

4. The fourth condition suggests that the process is both out of control and performing erratically.

In some cases it is desirable to plot the sample standard deviation values (using an S-chart) rather than the range values. This is done if we have reason to take advantage of the standard deviation's greater sensitivity to variation between data points within a sample. In either case, the S-chart and the R-chart both plot the variance within a sample.

## CHARTING INDIVIDUAL VALUES

By contrast to production processes—which can produce rapid streams of data in massive quantities—many service processes do not provide data at a sufficient rate to make it feasible to average several values to obtain a single data point on a control chart. In situations such as this, the length of time needed to obtain a single average from three or more measurements may expose the firm to problems that could go undetected in the meantime. Fortunately, SPC offers a solution for handling slow-moving data streams by using a control chart type referred to as a *chart of individuals*.

Let's say, for instance, that our law firm, Advocate General, is interested in tracking and controlling customer satisfaction as it relates to the handling of patent applications for its biotechnology clients, using data obtained from a survey instrument that rates customer satisfaction on a progressive scale of one to ten. If this information were available only every two or three weeks, the firm would likely obtain better insight by tracking this information with a chart of individuals rather than with an X-bar chart.

The control limits for the chart of individuals are typically determined from a simple calculation of something called the *average moving range* ($\overline{MR}$). This value is calculated by using a set of individual measurements that are collected over a period when the process is stable—in other words, during an interval when none of the values can be attributed to an anomaly. The centerline ($\overline{X}$) of the chart of individuals is determined by simply averaging the individual values over this same period of time, while the control limits (UCL and LCL) are determined by multiplying the $\overline{MR}$ value by 2.66 to obtain the three-sigma offsets on either side of the centerline. (An example showing the calculation of the $\overline{X}$ and $\overline{MR}$ values for a chart of individuals, using the data from our law firm application, is provided in Appendix B.)

Also, it should be noted that to establish "statistically valid" control limits for a chart of individuals, it is necessary that the data points exhibit a pattern of distribution that is reasonably close to a normal distribution. Recall from our discussion of the normal distribution in Chapter 7 that a quick test for "normalcy" can be performed by checking to see if approximately two-thirds (more precisely, 68%) of the data points fall within the region that is one sigma on each side of the centerline, with almost all of the remaining points falling in the regions that lie between one and two sigma on each side of the centerline.

In passing it might be noted that a chart of individuals without calculated control limits is called a *run chart*. Run charts are sometimes used to observe patterns of variation when we are not sure if valid control limits can be determined. Typically, however, a chart of individuals, with statistically valid control limits, is preferred over a run chart because the latter cannot tell us whether our process is in control or out of control.

## MONITORING ATTRIBUTE DATA

Now that we have seen how the X-bar and R-charts can be used to monitor variable data, we will focus the rest of this chapter on control charts used to monitor attribute data. Recall that attribute data are concerned with qualities that can only be counted or that can be identified as being present or absent. Attribute data, however, cannot be measured, at least in the sense that we would use special instruments to do so. Thus, attribute data are sometimes referred to as go/no-go data.

We should first point out that there are two types of attribute data: those that involve counting the number of defective items, and those that involve counting the number of individual defects (or flaws). This distinction is perhaps easiest to understand by way of

| Classification | Attribute to be monitored |
|---|---|
| Counting defectives | Percentage of rejections from the Patent Office |
| Counting defects | Number of mistakes on the patent application form |

illustration. Drawing once again on the example of the law firm, Advocate General, we can make the following distinctions:

Notice in the example corresponding to the first classification that a patent application can be either accepted or rejected by the Patent Office. If it is rejected, it is more accurate to say that the entire application is defective, rather than say that it has a certain number of defects. Also note, when speaking of the number of defectives it is equally valid to think in terms of the percentage of defectives, as with the metric used in this example. This is possible because we start with a known number of items, some of which are defective and the remainder of which are not.

The second classification in this example is predicated on the assumption that it is possible to examine the patent application and count the number of defects (also characterized as flaws or errors). In this case an entire application is not necessarily defective, but it is possible to count the number of individual defects. Be aware that in some cases a single flaw may be sufficient to deem an item defective. For instance, a missing signature on a patent application would likely be more critical than a grammatical error in the specification.

As far as defects are concerned, it doesn't make sense to speak of a percentage, because it may be impossible or impractical to determine the number of ways a defect might occur. For instance, there are numerous ways in which an error can be introduced on the patent application form. Thus we cannot say what would constitute 100 percent of all possible defects.

Why have we gone to so much trouble to distinguish between defects and defectives? We have done so because this distinction is

**Table 8.2** Selecting an attribute control chart.

| Classification | Sample Size | Control Chart |
|---|---|---|
| Defectives | Constant | P- or NP-Chart |
| Defectives | May Vary | P-Chart |
| Defects or Flaws | Constant | C- or U-Chart |
| Defects or Flaws | May Vary | U-Chart |

important from the standpoint of selecting the appropriate attribute control chart. Attributes representing a count of the number of defectives are plotted on the NP-chart (or the P-chart, in the case of percentage of defectives) while attributes representing a count of the number of defects are plotted on the C-chart or the U-chart.

Table 8.2 summarizes the conditions for selecting the appropriate attribute control chart. As you can see, selecting the right attribute control chart depends on whether our sample size remains constant or is allowed to vary from sample to sample.

Again, there is essentially no difference in the way we would interpret the patterns plotted by any of these control charts, even though, as we shall see, a different calculation is used in each case to determine the UCL and LCL values. The data values ultimately plotted on each chart will depend on the type of chart, as follows:

| Chart type | Data values plotted on chart |
|---|---|
| P-chart | Percentage of items (or units) that do not conform |
| NP-chart | Number of items (or units) that do not conform |
| C-chart | Number of defects (or flaws) detected per unit, assuming a fixed unit size |
| U-chart | Number of defects (or flaws) detected per sample |

It may help to consider a couple of examples to make it clear when a certain chart should be selected.

Let's say we are interested in tracking and controlling the number of typographical errors that find their way into various documents. Here we are talking about *counting* the number of defects, which calls for either a C-chart or a U-chart. If the documents vary in length, it would be appropriate to use a U-chart instead of a C-chart. In this case, for each document or portion of a document that is sampled, the data plotted on the U-chart might be the *number of typos per page*, or possibly, the number of *typos per thousand words*. If every document were the same length—as with the example of the patent application form—the C-chart would be appropriate. The centerline (average) in either case might be established by randomly selecting and examining previous documents, then eliminating those where problems due to *special causes* were known to have occurred. The calculation for the average value would be based on the average number of typos per page or per thousand words, in the case of the U-chart, or the average number of typos per application, in the case of the C-chart. These are represented by the symbols U and C, respectively.

As discussed earlier, a P-chart or NP-chart would be selected when it is necessary to judge an item (or unit) as being acceptable or unacceptable. If, for instance, we are interested in how many, out of some variable number of patent applications, are rejected by the Patent Office, we would compute the percentage of rejections in each case and then plot these as data points on a P-chart. On the other hand, if we are interested in how many rejections there are out of some fixed number of patent applications—let's say the number of rejects per 100 applications—we could simply count the number of rejects per 100 and then plot these as data points on an NP-chart. We might elect to establish the centerline for the P-chart by determining the average percentage of rejections based on historical data. If there were 18 rejections out of 500 applications in 2002, 24 rejections out of 550 applications in 2003, and 26 rejections out of 640 applications in 2004, we could compute the average percentage of rejections (or $\bar{P}$) as being 4.02 percent (that is, 68 rejections out of 1690 applications—which is *not* the same as averaging the averages for each year). We might elect to establish the centerline for the NP-chart (designated as $\overline{NP}$) by averaging the number of rejects per 100 applications, based, for example, on six sets of 100 randomly selected sets of applications over the past two years.

Establishing the UCL and LCL values for attribute data requires a different set of calculations from those used with variable data to construct the X-bar and R-charts. This stems from the fact that attribute data are not represented by a normal distribution because the data points are not equally distributed on either side of the centerline. The normal distribution, as you will recall from Chapter 7, is characterized by the bell-shaped curve, which allows for an equal number of values on either side of the centerline (or average value). Attribute data do not typically distribute in this manner.

Without going into detail, we will leave this chapter by pointing out that the UCL and LCL values for the attribute charts can be computed by using the formulas shown in Table 8.3.

Note: The lower control limit on an attribute chart will never be less than zero because you can never have fewer than zero defects or defectives. If the calculated value for the LCL is less than zero, simply show zero as the LCL on the control chart. Also note that $\bar{P}$ is

**Table 8.3** Formulas for UCL and LCL on attribute charts.

| Chart | Average | Standard development | Upper control limit | Lower control limit | Comments |
|---|---|---|---|---|---|
| P | $\bar{P}$ | $\sqrt{\bar{P}(1-\bar{P})/n}$ | $\bar{P}+3s$ | $\bar{P}-3s$ | n is the average sample size used in establishing the control limits |
| NP | $N\bar{P}$ | $\sqrt{N\bar{P}(1-\bar{P})}$ | $N\bar{P}+3s$ | $N\bar{P}-3s$ | P-bar is a decimal amount for both the P-chart and NP-chart calculations |
| C | $\bar{C}$ | $\sqrt{\bar{C}}$ | $\bar{C}+3s$ | $\bar{C}-3s$ | |
| U | $\bar{U}$ | $\sqrt{\bar{U}/n}$ | $\bar{U}$ | $\bar{U}-3s$ | n is the average sample size used in establishing the control limits |

often represented in lowercase, so that $\bar{P}$ becomes $\bar{p}$ and $N\bar{P}$ becomes $N\bar{p}$.

# 9 | Process Capability and Control Limits

We have been careful to stress the point that the control limits and the centerline (that is, average) value should be established on the basis of what we know to be normal operating conditions. In other words, we need to determine what the process in question is capable of performing, rather than how we would like for it to perform. It is entirely possible that the process, as it exists, is not capable of performing the way we would wish. Under these conditions, there could conceivably be occasions when the process would be in control, as far as the SPC charts are concerned, but not in conformance with the specification requirements.

With this in mind, we can distinguish between control limits and specification limits. The control limits speak to the actual capability of the process, and the specification limits speak to the minimum desirable standards.

Though it is possible that the specification limits have been established arbitrarily, it is more likely that they have been based on certain rational criteria, such as achieving a competitive advantage, customer satisfaction, or minimum reliability standards. In other words, the specification limits are typically management imposed, rather than being statistically determined. As with the control limits, the specification limits are designated as USL (upper specification limit) and LSL (lower specification limit).

The following point bears repeating to ensure there is no confusion between *control limits* (which are typically calculated from sample data) and *specification limits* (which are mandated, perhaps by the customer):

> The SPC control limits (UCL and LCL) should be predicated on actual process capability, rather than desirable specification limits—if, that is, we are truly interested in statistically based control limits.

The *baseline* operating conditions that help us define our *process capability* can be established in two ways:

1. By examining historical data that are randomly selected specifically for this purpose, or
2. By designating a starting point to observe the process in action while collecting data that can be used to develop the control limits

In either case, the following two factors should be taken into consideration:

1. Enough data should be collected to represent the process over its full operating range.
2. Data corresponding to problems resulting from special causes should be eliminated from the data set used to establish the control limits; otherwise, the control limits will be predicated on an unstable process.

Perhaps it is clear by now that our process control limits will typically be *tighter* than our specification limits—at least when we are plotting sample averages (X-bar values). In reality, it is the spread (or distribution) of the individual values in relation to the specification limits, rather than the control limits per se, that concerns us. Although the control limits are predicated on the distribution of the individual measurements, a simple mathematical relationship between the two allows us to calculate one, given the other. In other words, if we are starting with known values for the control limits—as we might if we were examining a pre-existing control chart—we can determine the plus and minus three-sigma values (or spread) of the distribution of individual measurements that gave us the UCL and LCL values in the first place. Let's see how this works.

Without getting into a theoretical discussion of why, for a control chart that is used to plot sample averages (for instance, the X-bar chart), each control limit (UCL and LCL) should be no further away from the average (centerline) than an amount equal to the USL (or LSL) divided by the square root of the sample size ( $\sqrt{n}$ ). In other words, the UCL should show up on the chart as being less than USL ÷ $\sqrt{n}$ . Likewise, the LCL should show up on the chart as being greater than LSL ÷ $\sqrt{n}$ . (For the other chart types described in Chapter 8, the specification limits should be offset from the average value by an amount equal to at least three times the standard deviation value that was used to construct the control chart.) For convenience, these computations are repeated in Table 9.1.

**Table 9.1** Specification limits versus control limits.

| Chart Type | Desirable Condition |
|---|---|
| X-bar Chart | UCL < USL ÷ $\sqrt{n}$ <br> or <br> USL > UCL × $\sqrt{n}$ <br> LCL > LSL ÷ $\sqrt{n}$ <br> or <br> LSL > LCL × $\sqrt{n}$ |
| Attribute Charts | UCL < USL <br> or <br> USL > Average + 3s <br> LCL > LSL <br> or <br> LSL < Average − 3s |

Let's return to our law firm, Advocate General, to see what all of this means in more tangible terms.

The control point designated as *Delay in Passing the Patent Application Through the Review Committee* was described in Chapter 3 as requiring a measurement of time, making it possible to construct an X-bar chart. Starting with the control limits given in Figure 8.1 (UCL = 32 and LCL = 16) and the assumption that the firm wishes to *guarantee* its clients that a patent application will not require more than 50 hours to review, let's see whether the upper control limit is sufficiently *tight* to stand up to this goal. In other words, we need to determine whether the UCL is in fact less than the USL divided by the square root of the sample size. We see that:

1. Sample size = 3, and the square root of 3 = $\sqrt{3}$, or 1.73.

2. UCL should be less than 50 ÷ 1.73.

3. Because 50 ÷ 1.73 is 28.9, and our UCL value is 32, the evidence *does not* support the claim that a patent application won't require more than 50 hours to review, at least occasionally. (Quality professionals would label this process *not capable*.)

As this example shows, it is important to keep in mind that the process control limits are determined on the basis of how the process is capable of performing in light of current conditions. We cannot arbitrarily change the control limits on the basis of what our specification limits are set at—that is, if we are truly interested in maintaining statistical-based control limits. Instead, we must improve the long-term performance of the process—as earlier alluded to and further discussed in Chapter 11—in order to *tighten* the distribution of individual values and therefore tighten the control limits, which are a function of the individual values. If Advocate General is serious about standing behind the claim that no patent application will require more than 50 hours to review, it must take some permanent action to improve the process.

For now, simply bear in mind that if a process is modified—whether for good or for bad—we will likely have to reestablish the control limits and the average value. In this case, historical data will be of no value in establishing either the new control limits or the new average value. Also bear in mind what we said regarding overregulation; it is typically unwise to adjust the process simply on the basis of the latest sample results.

In Chapter 11 we will revisit the issue of process spread in the context of improving the process.

# 10 Collecting and Plotting Your Data

Our discussion has mostly been concerned with knowing what needs to be done to set up a statistical process control subsystem and seeing what this powerful tool can accomplish. In this chapter we will turn our attention toward a set of issues concerning *how* and *who*.

We specifically want to say a few words about the following issues, which in turn give rise to a set of questions that require organization-specific answers:

- How and where the SPC data are collected
- Who will be responsible for collecting the data and maintaining the charts
- How often to collect data
- Who inspects and/or reviews the results
- Who is responsible for taking corrective action

Because the human element is firmly embedded in these issues, they often involve complexities that are fueled by individual quirks such as personality differences, biases, and hidden agendas. Furthermore, institutional factors such as the culture of the organization, the general attitude toward change, the level of trust between upper management and the rank and file, and the degree to which the process requires cross-functional cooperation, may also need to be taken into account. All of this simply underscores the fact that a willing and able leader must be involved from the outset and that the SPC implementation process must be planned and managed with the same diligence that would be applied to any significant organizational change initiative.

Rather than attempting to address intricacies that are workplace dependent, we will speak to the primary concerns that are universally relevant to setting up and maintaining an SPC subsystem.

Questions relating to each of these issues are provided in this chapter in the form of checklists, to assist in deriving answers that apply to your situation. Also, for clarification, a brief statement of intent has been provided to expand on the rationale behind the question.

## WHERE AND HOW SHOULD DATA BE COLLECTED?

The issue of determining where and how the SPC data will be collected leads to deeper questions that must account for the fact that certain aspects of the process are more critical than others and that differences in opinion may exist on how best to measure and monitor process performance. Answers to these questions should be laid to rest well before jumping into a process capability study that pertains to a particular process performance indicator.

Though obvious in some cases, answers to the following will need to be discerned:

_____ Which elements of the process are critical in terms of their impact on the criteria by which process performance is judged?

*Intent: To determine which components of the process are primary candidates for being monitored and controlled*

_____ What are the possible indicators of performance at each point in the process that we wish to monitor and control?

*Intent: To pinpoint key quality characteristics that can be translated into metrics that may be monitored and controlled by using SPC*

_____ What, if any, data collection forms are needed?

*Intent: To devise any special forms that will serve as a convenience in collecting and organizing raw data—such as tally sheets*

_____ What, if any, special instruments are needed to quantify and measure the various indicators of process performance?

*Intent: To devise any special instruments that may be needed to obtain quantitative data indicative of process performance—such as a customer-satisfaction instrument with a ten-point rating scale*

\_\_\_\_ Are special skills needed to measure the various performance indicators?

*Intent: To determine whether the individuals responsible for collecting the data will need special training in using the measuring instruments or taking measurements without otherwise disturbing the process*

\_\_\_\_ What, if any, negative influence will the data collection activities have on the process itself?

*Intent: To anticipate and minimize the risks associated with collecting certain types of data, including ways in which the data may be manipulated or abused—such as shortcutting those aspects of the process that are not charted*

\_\_\_\_ How will historical records be maintained and by whom?

*Intent: To identify how SPC archival data will be stored and retrieved for later use and who will bear this responsibility*

## WHO SHOULD BE RESPONSIBLE?

Typically the person or persons most intimately involved with the process at the point we wish to monitor and control will bear the responsibility of collecting data and maintaining the control charts. Nevertheless, there may be valid reasons for doing otherwise. It is wise to ensure that these responsibilities are clearly understood and agreed upon.

With this in mind, we may need to determine the following:

\_\_\_\_ Who will collect data at the points we wish to monitor and control?

*Intent: To identify the individuals by name who will bear primary, and perhaps backup, responsibility for collecting data—being careful to avoid gaps or overlaps of responsibility*

____ Has the time impact resulting from these activities been factored into the workload?

*Intent: To establish reasonable estimates of the time it will take to collect data and maintain the control charts, taking action to redistribute workloads if necessary*

____ Have the responsible parties been adequately trained in the SPC fundamentals that pertain to their roles and responsibilities?

*Intent: To clarify the knowledge needs of the individuals who will maintain the SPC subsystem—or use the information that it generates—and provide training to the degree necessary*

## HOW OFTEN SHOULD DATA BE COLLECTED?

The issue regarding how often to collect data is influenced by two factors: the minimum frequency necessary for detecting potential problems, and practical considerations concerning the time and cost involved in collecting data.

Even if time and cost were not factors, for the sake of making sense out of raw data it is often more desirable to use sampling techniques to produce summary statistics rather than recording every possible data point.

Although there are no hard and fast rules regarding the sampling frequency, one approach calls for taking samples more frequently in the beginning, at least until the process is proven to be stable, and then reducing the sampling frequency. Certainly, any process that is subject to recurring problems should be monitored more closely than otherwise. Furthermore, the cycle time of the process should be taken into consideration. For instance, the time between samples would be shorter for a process having a cycle time of minutes compared to one having a cycle time of days. Care should be taken not to inject bias into the sampling process.

Collecting and Plotting Your Data    81

With these factors in mind, questions related to this issue might include:

_____ What do we know about the stability of the process, in particular each component of the process that we intend to monitor and control?

*Intent: To determine whether the process, at each point of interest, is capable of providing data that will allow for distinguishing between variation due to special causes and variation due to common causes*

_____ How often should we collect data in order to feel confident that problems occurring between samples will not go undetected?

*Intent: To establish the sampling frequency for each data collection point that is consistent with the rapidity with which an out-of-control condition could arise*

_____ Are the necessary human and financial resources available to support the desired sampling frequency? If not, what trade-offs are necessary?

*Intent: To secure additional resources if needed and generate options if additional resources are needed but not available*

_____ What influence might the data collection activities have on the process itself?

*Intent: To determine whether and how the data collection activities will interfere with the operation of the process—for instance, a slowdown in the process to collect data from the people supporting that process*

## WHO INSPECTS OR REVIEWS THE RESULTS?

Quite often someone other than the person or persons responsible for collecting the data and maintaining the control charts will serve as an *auditor* or will at least be responsible for occasionally reviewing the control charts. This is done not to question the integrity of the person doing the data recording, but rather to ensure that problem conditions are not being overlooked or that significant trends are not being missed.

Questions related to this issue might include the following:

_____ How often should each control chart be reviewed?

*Intent: To ensure that problems are being identified accurately and that an alert for action is conveyed up the chain of command if necessary*

_____ Who is the logical choice for reviewing each of the various control charts?

*Intent: To select the reviewers based on their knowledge of the process, their grasp of SPC fundamentals, and their authority to make decisions pertaining to the process*

_____ How will trends and other potential problem conditions be flagged?

*Intent: To establish the criteria for judging the criticality of a potential problem and to specify the method for alerting others of a problem condition based on the immediacy for action*

Note: As indicated by the example in Figure 10.1, it is common practice to make brief notes on the control chart indicating awareness of a potential problem, the possible explanation for why it exists, and/or what corrective action was taken or recommended.

## WHO TAKES CORRECTIVE ACTION?

When a problem condition is detected or a suspicious trend is observed, it should be crystal clear who has responsibility for bringing the process back in control—or in the case of a trend, ensuring that it does not drift out of control. This responsibility should be established in the beginning, rather than waiting for a problem to occur and then trying to determine who has responsibility in the heat of the moment.

While this issue is rather straightforward, some related factors may need to be considered:

_____ What authority will the person who maintains a particular control chart be given with regard to taking corrective action in response to a problem?

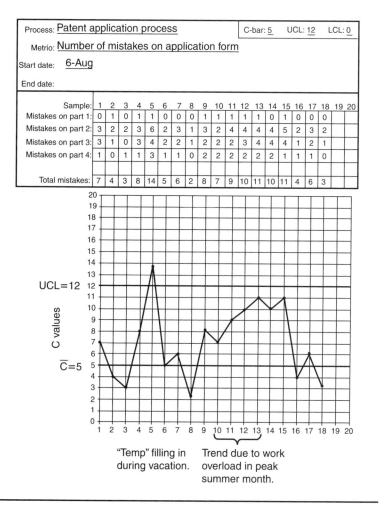

Figure 10.1 C-chart with comments added.

*Intent: To specify the limits of the problem-solving and decision-making authority that will be invested in the individuals who are responsible for maintaining the various control charts*

_____ If not the data recorder, then who is responsible for taking corrective action?

*Intent: To specify who has responsibility for action in dealing with problems that require escalation*

____ How will any corrective action be verified and the results documented?

*Intent: Depending on the criticality of the process, it may be desirable to designate individuals who have "sign-off authority" to confirm that any corrective action has had the desired effect*

____ Who will be responsible for making a problem impact assessment?

*Intent: Specifies the individual or individuals who are capable of translating process problems into costs to the organization or impact on the customer, and communicating the potential consequences of a no-action decision*

Issues related to the *how* and *who* of collecting and plotting the SPC data can be emotionally charged, making them some of the most difficult to deal with. The best defense, from a planner's point of view, is to anticipate the concerns likely to arise and be prepared to address—not dismiss—these concerns. There will be workplace-specific issues, for sure, but the questions posed in this chapter are a good place to start.

# 11 Continuous Improvement

This chapter examines several issues related to process improvement to shed light on how SPC relates to this important subject. Continuous improvement is a never-ending quest—one that may eventually lead to a radical makeover of the process. More often than not it will involve certain planned projects that are designed to achieve incremental improvements, moving us ever closer to the ideal state of the process in question.

As you may recall, it was mentioned earlier that if we wish to *tighten* our statistical process control limits (that is, the UCL and LCL), something will have to be done to improve the long-term performance of the process, rather than simply eliminating occasional problems. We also noted that the average value and the control limits for each of the various SPC charts depend on *current* process capability—in other words, its performance capability given its current design and operational limitations.

In a broader sense, however, we are not interested in simply tightening the control limits as an end unto itself, but with improving the consistency and stability of the process as well as the primary performance indicator, such as the *Time Required to Process a Patent Application*. As we have seen, the control limits simply reflect how the process is performing and are typically calculated from actual data.

The notion of improving the consistency and stability of the process, as well as its *on-average* performance, can be visualized with the aid of Figure 11.1.

If the example shown in Figure 11.1 indicates the amount of time it takes our law firm time to process a patent application within two standard deviations of the average, then we can deduce the following:

Before improvements were made:

- On average it took 24 hours to process a patent application.

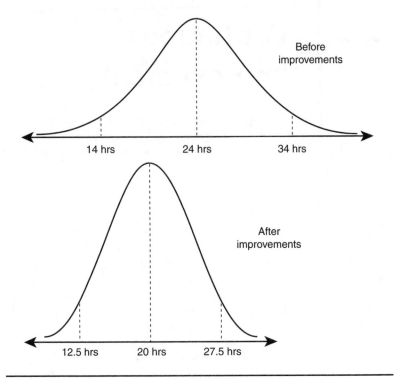

**Figure 11.1** Before and after improvements.

- 95 applications out of 100 were processed within 14 to 34 hours—corresponding to a time span of plus or minus 2 standard deviations on each side of the 24-hour average (based on the characteristics of the normal distribution, described in Chapter 7).

After improvements were made:

- On average it now takes 20 hours to process a patent application.

- There is a 95 percent chance that an application can now be processed within 12.5 to 27.5 hours, a span of 15 hours, compared to 20 hours previously.

In this example, the improvements resulted in both a reduction in the *average* time to process a patent application and more *consistency* in the time it takes to process an application. These will be reflected on the control chart in the form of a lowered centerline and narrower control limits, respectively.

It follows that if we wish to improve the overall performance of a certain process, one or both of the following actions will have to be taken:

1. The *common causes* of variance will have to be identified and the process will have to be modified or improved to minimize the degree of variation that is contributing to the wide swings in process performance. Recall that in the case of attributes, the control limits can be tightened by decreasing the average number of defects or the average number, or percentage, of defectives—depending on the particular attribute control chart.

2. The process will need to be examined to determine how the average value that corresponds to a particular control point can be modified. The process itself will have to be overhauled accordingly. (Note: In the case of a parameter such as process cycle time, it is desirable to decrease the average value, while in the case of a parameter such as *Customer Satisfaction Rating* it is desirable to increase the average value.)

Quite often, both forms of improvement are desired, as in the case of our patent application process.

Again, we stress that the control limits and the average value will need to be reestablished if the process has been permanently modified. Using the previous example, after the process is improved, a new process capability study may show that both forms of improvement described above have been realized and that we need to revise Figure 8.1 so that the average value is now 20 and the UCL is 26.5 and the LCL is 13.5. In other words, our process capability study indicates that, on average, we can expect a patent application review to require 20 hours and that 99.7 percent of the *sample averages* (not individual items) will fall within 6.5 hours of this time based on a sample size of 3. (Recall that a distribution of sample averages displays the characteristics of a normal distribution; therefore, we know that 99.7 percent of the values will fall between LCL and UCL.)

In examining a process for possible improvements, it is important to consider the *interdependent* nature of the various subprocesses. If, for example, we wish to decrease the overall cycle time of a process, it would make little sense to improve some part of the process that would only result in a backlog of work further downstream. Care must also be taken to ensure that by optimizing a certain subprocess we do not inadvertently cause suboptimum performance elsewhere within the same, or perhaps another, process. This is particularly likely if processes or subprocesses cut across organizational boundaries.

Ultimately we may discover that a certain process needs to be changed significantly to realize the degree of improvement we desire. In the early 1990s, the term *process reengineering* came into fashion as a strategy and mind-set for making radical changes to certain critical processes within an organization. Though the term has fallen out of favor in the United States because of misuses and abuses that have occurred under the banner of process reengineering, such is not the case in Japan and other Asian countries.

Those who ascribe to the process reengineering philosophy—or process redesign, if you prefer—believe the organization should be changed, however necessary, to fit the processes, rather than vice versa. Not only is each process examined in detail, but the relationships between the various processes also are examined. Considering again the case involving the law firm, Advocate General, it may be found that two of the firm's three major processes, Contracts and Labor Law, have certain subprocesses in common, possibly highlighting the opportunity to consolidate these subprocesses for improved efficiency. Incremental improvements in either of these two processes in isolation would not likely have identified this opportunity.

Though process reengineering may start with a fresh look at the way a certain process is structured and executed, a less dramatic approach is to begin with an examination of the value-added and non-value-added aspects of existing processes to identify substantial improvement opportunities, sometimes using a technique known as *value analysis*. Elements or subprocesses that add value to the process from an internal or external customer's perspective are enhanced, and those that add cost without adding value are diminished. Tools such as Pareto analysis (discussed in Chapter 6) and cost-benefit analysis may assist in making this assessment. As with subprocesses, care must be taken to ensure that optimizing the performance of one process does not lead to suboptimum performance of another—a likely possibility when optimization involves shifting resources.

In contrast to process reengineering—which is done as needed—many companies are committed to the philosophy that quality improvement is a never-ending process. These companies are often driven by a self-imposed goal of virtually eliminating defects or achieving some other ideal, such as consistently delivering products and services on time. Although *breakthrough improvements* in quality are desirable, these companies encourage the pursuit of *incremental improvements* as well. In reality, it should be noted that reengineering and continuous improvement are not in competition. In fact, continuous improvement can pick up where reengineering leaves off by continuing the cycle of improvement until radical changes are called for.

We should also note that, technically speaking, the notion of defect elimination refers to attribute data, because by definition attributes result from counting the number of defects or counting the number (or percentage) of defectives. For instance, it is conceivable that our law firm, Advocate General, might wish to establish a goal of generating error-free patent applications, which involves counting the number of errors on the patent application form. Informally, however, the notion of eliminating defects may also be applied to the target value for certain variables. Any deviation from the target would be considered a loss of some degree. Furthermore, variables *within* the process may be monitored and controlled with the ultimate intent of reducing the number of defective items in the end—making it a part of an overall defect-elimination policy.

What, you may ask, does all of this mean to the individual contributor? Simply this: Whether the emphasis is on process reengineering or continuous, incremental improvements, everyone who has a role in support of the process will likely be involved in suggesting, implementing, or verifying process improvements. And, as we have stressed, when improvements are implemented, the process capability will have to be reestablished. This is a fact that everyone associated with the process should be aware of, especially if decisions regarding corrective and preventive action will be based on the control chart results.

For the reasons discussed in this chapter and throughout this book, statistical process control is often the method of choice for validating the performance of a process by objectively confirming (or disavowing) claims that performance has actually improved in response to incremental changes or major modifications. Coupled with its use in solving and preventing problems, this makes SPC an invaluable tool for everyone who has a role in supporting and improving processes within the organization.

# Concluding Remarks

While there is obviously much more that could have been said in this book on the subject of statistical process control, some difficult decisions had to be made regarding what to include and not include, as well as the "angle" in which the material was presented—the *side door*, if you will. The reader is reminded that our primary goal was to make this important topic accessible and user friendly to the target audience: right-brain thinkers who are perhaps new to the subject, rather than statisticians or seasoned quality professionals. Nevertheless, we encourage anyone who wishes to learn more about SPC to delve deeper—perhaps by taking a company training class, by embarking on a self-study program, or by becoming actively involved in a professional association in which learning opportunities abound, such as the American Society for Quality (ASQ).

We leave you now with an important fact and hope it will inspire you to continue what you have started by reading this book. Though SPC may not qualify as "the latest and greatest thing," one thing is for sure: *It works!*

# Appendix A
# Constructing a Pareto Analysis Chart

Pareto analysis was introduced in Chapter 6 as a tool for identifying the causes of problems within a process and ranking these problems according to their frequency of occurrence. Once identified, it may be desirable to establish metrics related to the most significant causal factors and then track these metrics with the aid of control charts. From a process control perspective, it is especially desirable if the metric in question can serve as an early warning indicator of an impending problem.

Referring to Table A.1 and Figure 6.3 on page 50 by way of example, here is the step-by-step procedure for constructing a Pareto analysis chart in an electronic spreadsheet:

1. For each problem that poses a concern, identify as many causes as possible—perhaps by using brainstorming or the fishbone diagramming technique.

2. For each problem cause of interest, determine the cause's frequency and the grand total.

3. Following the format of Table A.1, list the problem causes in order according to their frequency (highest to lowest).

4. For each cell in the percent of total column, calculate the percentage of the total that each problem cause represents by dividing its frequency of occurrence by the grand total and then multiplying this value times 100 to obtain a percentage.

5. Starting with the problem cause at the top of the list, calculate the value for each cell in the cumulative frequency column (column 5) by adding the frequency of occurrence of the event on the line where the cursor is

**Table A.1** Data used in constructing the Pareto analysis chart.

| | Errors in patent applications submitted to Patent Office | | | | |
|---|---|---|---|---|---|
| Item | Source of Error | Frequency | % of Total | Cumulative Frequency | Cumulative Percent |
| 1 | Incomplete Specifications | 98 | 32.3% | 98 | 32.3% |
| 2 | Drawing errors | 74 | 24.4% | 172 | 56.8% |
| 3 | Invalid claims | 46 | 15.2% | 218 | 71.9% |
| 4 | Submission errors | 30 | 9.9% | 248 | 81.8% |
| 5 | Cross-reference errors | 19 | 6.3% | 267 | 88.1% |
| 6 | Formatting errors | 12 | 4.0% | 279 | 92.1% |
| 7 | Filing fee errors | 9 | 3.0% | 288 | 95.0% |
| 8 | Patent search misses | 6 | 2.0% | 294 | 97.0% |
| 9 | Omitted signatures | 5 | 1.7% | 299 | 98.7% |
| 10 | Other | 4 | 1.3% | 303 | 100% |
| **Total** | | **303** | **100%** | | |

positioned to the frequency of occurrence of each event that precedes it in the list.

6. Using the ordered data from the table from the top down, create a bar chart depicting the frequency of occurrence of each problem cause (as indicated on the vertical scale on the left side of the chart).

7. Plot the cumulative percentage curve on the chart, using the calculated values from the table and the vertical scale on the right side of the chart that spans from zero to 100 percent.

Note: In some cases it may be desirable to rank problem causes by using a criterion other than frequency of occurrence—such as the likelihood (or probability) that each causal factor, from a set of factors, will trigger a certain risk event. In such cases, the Pareto principle may still apply, at least in the sense that a few of the causal factors are much more critical than the others. Be aware, however, that the cumulative percentage calculation will be meaningless, as it is in this example, if the ranking criterion you are using does not yield values that can be logically summed. If, for instance, there is an 80 percent chance a customer will defect because of slow response time and a 40 percent chance a customer will defect because of a billing error, it is meaningless to add these values, because they are independent measures of the stand-alone effect that each factor has on customer loyalty.

# Appendix B
# Charting Individual Values

The basis for using a chart of individuals was described in Chapter 8. Table B.1 demonstrates how the average moving range ($\overline{MR}$) and the centerline ($\overline{X}$) for this chart are calculated, using a set of measurement data from the example involving our law firm, Advocate General. This example involves an infrequent measurement of customer satisfaction in handling the patent applications on behalf of the firm's biotechnology clients. This measurement can lie anywhere on a scale of one to ten, where ten represents total customer satisfaction.

Refer to Table B.1 and observe the following:

1. The centerline value ($\overline{X}$) was calculated by averaging the individual values.

2. The moving range corresponding to a particular data item is simply the magnitude of the difference between the values of two sequential values, starting from the top down. For instance, for item 3 the moving range is the unsigned difference between 5 and 6, and for item 7 the moving range is the unsigned difference between 8 and 5.

3. The average moving range ($\overline{MR}$) is calculated by averaging the unsigned difference values in the moving range column. Notice, there will always be one less moving range value than there are data values, so one less value is used in calculating the average moving range than in calculating the centerline value ($\overline{X}$).

**Table B.1** Calculating the moving range and centerline values.

| Item | Date | Value | Moving range |
|---|---|---|---|
| 1 | 12 June | 7 | |
| 2 | 20 June | 6 | 1 |
| 3 | 14 July | 5 | 1 |
| 4 | 21 July | 7 | 2 |
| 5 | 26 July | 7 | 0 |
| 6 | 17 August | 5 | 2 |
| 7 | 19 August | 8 | 3 |
| 8 | 30 August | 6 | 2 |
| 9 | 11 September | 5 | 1 |
| 10 | 27 September | 6 | 1 |
| Centerline: | | **6.2** | |
| Average moving range: | | | **1.44** |

The control limits, based on these data, may be computed as follows:

$$\text{UCL} = \bar{X} + 3\sigma \quad \text{or} \quad \text{UCL} = \bar{X} + 2.66\,\overline{\text{MR}} = 6.2 + (2.66 \times 1.44) = 10.03$$

$$\text{LCL} = \bar{X} - 3\sigma \quad \text{or} \quad \text{LCL} = \bar{X} - 2.66\,\overline{\text{MR}} = 6.2 - (2.66 \times 1.44) = 2.37$$

Note: If the calculated value for either control limit exceeds the limits of the scale—as it does in this case for the UCL—simply set the control limit at the maximum or minimum value possible, whichever applies.

## Charting Individual Values 97

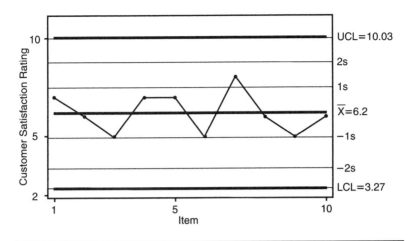

**Figure B.1** Chart of individuals.

Figure B.1 shows how the data and control limits from this example would appear, plotted with the aid of the statistical software package MINITAB®.

# Glossary

**assignable cause**—A problem cause arising from an anomaly that manifests itself as variation that cannot be attributed to the inherent characteristics of the process. Also called a *special cause*.

**attribute data**—A data classification concerned with qualitative characteristics, as opposed to measurable characteristics.

**average chart**—An SPC control chart that plots the *average value* of samples taken from a certain *population*. Also called an *X-bar chart*.

**average moving range**—A value used in establishing the control limits on a *chart of individuals,* which is computed by averaging the magnitude of the difference between adjacent data points in a data set.

**average value**—A value obtained by adding a set of numbers and then dividing this sum by the number of items in the set.

**baseline value**—The initial value of a certain quantifiable characteristic of a process that is often used as a basis for comparing future values of that same characteristic.

**bell-shaped curve**—See *normal distribution.*

**best estimate**—An estimate of a single value from a set of data based on the average value of that data set.

**breakthrough improvements**—Major improvements made by rethinking and redesigning the structure of a process and/or the way it operates. Often undertaken to improve competitiveness.

**C-chart**—An SPC control chart that plots attribute data in the form of number of defects counted when the sample size is fixed.

**centerline value**—The reference line between the upper and lower control limits of a control chart; it is calculated by averaging a set of data values that are collected under controlled conditions.

**chart of individuals**—A type of SPC chart for plotting individual values, as opposed to the average values of sample groups.

**common cause**—A cause of variation inherent within a process because of the way the process is designed and used, rather than by an abnormal problem condition.

**continuous improvement**—An organizational philosophy based on a mind-set and action-orientation that seeks to continuously improve processes so as to eliminate waste and better serve customers.

**control chart**—Any of the various SPC charts that track process performance.

**control limits**—The upper and lower boundaries on a control chart; exceeding them suggests the need for corrective action. (See *upper control limit* and *lower control limit*.)

**critical to quality (CTQ)**—A qualitative or quantitative outcome characteristic of a certain process or product that is essential from the customer's perspective.

**CuSum chart**—An SPC chart that tracks the cumulative difference between the *average value*s of our samples and the baseline average of the control chart.

**cycling pattern**—An unnatural pattern of data points on a control chart that appears as a repetitive cycle rather than a random pattern of variation.

**defects per million opportunities**—The number of errors or defects, out of a million possible occurrences, that are found over a specified period in a certain product or process.

**dispersion**—The degree to which a set of data points are spread out relative to the average value; it is often expressed as a single number, such as the standard deviation.

**distribution**—An ordered display of a set of data values that shows the frequency with which each value (or narrow range of values) occurs or the likelihood that each value will occur.

**firefighting**—A colloquial term referring to a reactive, as opposed to a preventive form of problem solving, often caused by poor planning or mismanagement of priorities.

**grand average or grand mean**—The centerline value of an X-bar chart; it is calculated by averaging a set of sample averages.

**in-control process**—A desirable condition when the process is tracking within its SPC control limits and showing no indication of an impending problem.

**incremental improvements**—A series of small-scale improvements that are made in a process over time, often with the intent of reducing waste and/or decreasing process cycle time.

**lower control limit (LCL)**—The lower boundary on an SPC control chart, typically set at three *standard deviations* below the average. May be set to zero if the computed value of the LCL is a negative number.

**lower specification limit (LSL)**—The low-end specification value established on the basis of how we desire the process to perform, rather than how it is truly capable of performing.

**magnitude**—The numerical value of a certain quantity, irrespective of its polarity; for instance, the magnitude of $-8$ is the same as that of $+8$.

**mean**—Same as *average value*.

**mixture pattern**—A pattern of variation on a control chart that results from mixing data from two distinct aspects of a process; for instance, a composite plot of the time required to process two types of patent applications.

**natural systems**—Systems that occur in nature. The behavior of such systems often displays a pattern of variation that plots as a *normal distribution*.

**non-value-added activities**—Any steps or activities in a certain process that do not contribute to the functional intent of that process.

**normal distribution**—A distribution of data points that assumes a symmetrical bell shape, with the peak of the curve occurring at the average value.

**NP-chart**—An SPC control chart that plots *attributes* in the form of *number of defectives*.

**number of defectives**—The number of items or units in a sample that do not meet the pass/fail criterion.

**number of defects**—The number of flaws or *defects* an item or unit may possess, such as spelling errors on a patent application form. In the case of minor defects, a single defect may not be sufficient to render an entire unit defective.

**out-of-control process**—An undesirable condition in which the process is tracking outside of its SPC control limits or is exhibiting a pattern that might indicate an impending problem.

**P-chart**—An SPC control chart that plots *attributes* in the form of percent defective.

**Pareto analysis**—A graphical technique for plotting the sources of a particular problem, ranked in relation to their overall impact. Isolates the significant few causes from the insignificant many, allowing for a focused process improvement effort.

**population**—The entire set of items, values, or *variables* that make up a certain group of factors we may be interested in measuring or counting.

**probability distribution**—A distribution displaying the likelihood (or probability) of occurrence for each value in a given data set.

**process**—The inputs, outputs, customer expectations, work content elements, and control mechanisms (such as SPC) that are linked by a common objective of transforming customer needs into solutions.

**process capability**—What the process, as it is designed and operated, is capable of producing, as opposed to what we may desire. Because of its design, a process could be in control but incapable of staying within its tolerance limits.

**process control**—A set of activities that ensure that a certain process behaves as we would expect, given the current capabilities and limitations of its design.

**process management**—A philosophy that views the organization as a set of interrelated processes, structuring it to best support these processes.

**process reengineering**—An approach to analyzing and designing processes that calls for a radical overhaul of methods, rather than machines, by first determining the most efficient method of operation and then the technology needed to implement it.

**quality characteristic**—A certain process variable or attribute we wish to monitor by using SPC. Also called a *performance indicator*.

**R-chart**—An SPC control chart that plots *variables* in the form of range values.

**reengineering**—See *process reengineering*.

**sample**—A set of randomly selected data items taken from a certain *population*.

**sigma**—Greek letter $\sigma$ used as a shorthand representation for the *standard deviation* of a *population*.

**Six Sigma**—A customer-focused approach to quality predicated on improving a certain process or product so that errors or defects will occur no more than 3.4 times out of a million opportunities.

**special cause**—An induced, rather than inherent, source of variation within a process. Commonly identified by an unnatural pattern of variation on an SPC control chart. Also called an *assignable cause*.

**specification limit**—A specified value established on the basis of how we desire the process to perform, rather than how it is truly capable of performing. (See *upper specification limit* and *lower specification limit*.)

**stability condition**—The state of a certain process during an interval when there are no known problems due to *special causes*.

**standard deviation**—A calculated index of the degree of variation, or dispersion, that exists within a certain *population*.

**statistic**—Some quantifiable characteristic of a sample.

**statistical process control (SPC)**—The application of statistical techniques to monitor process performance and maintain process control.

**statistical quality control (SQC)**—A term often used interchangeably with *statistical process control*, though technically it encompasses acceptance sampling as well as SPC.

**statistically valid**—Refers to control limits (on an SPC chart) that are predicated on actual, quantifiable characteristics of the process rather than guesses or wishful thinking.

**stratification pattern**—An unnatural pattern on a control chart that appears as a series of data points that hug the centerline rather than displaying a pattern that is consistent with the distribution of data points based on what we expect.

**symptom indicator**—An observable or measurable characteristic of some underlying condition, such as job errors as an indicator of mental fatigue.

**trending condition**—An unnatural pattern on a control chart that appears as a series of increasing or decreasing data points.

**U-chart**—An SPC chart that plots attributes in the form of number of defects counted, without requiring a fixed sample size.

**unnatural pattern**—The pattern of a set of data points plotted on a control chart that suggests the variation within a certain process is caused by something other than *common causes*.

**unstable process**—A process in which the centerline and/or the distribution of the data points of a certain variable do not display a consistent pattern.

**upper control limit (UCL)**—The upper boundary on an SPC control chart, typically set at three standard deviations above the average.

**upper specification limit (USL)**—The high-end specification value established on the basis of how we desire the process to perform, rather than how it is truly capable of performing.

**value-added activities**—Any of the steps or activities in a certain process that are critical in fulfilling the functional intent of that process.

**value analysis**—A technique used to identify the basic value elements of a process with the intent of minimizing anything that does not add value to that process.

**variable**—A quantifiable characteristic of a process or product that can assume different values at different times as a result of special and/or common causes of variation.

**variable data**—A data classification concerned with certain quantitative characteristics that can be measured.

**X-bar chart**—See *average chart.*

**zero defects**—A goal of continually improving an organization's processes to the point where every output conforms to specification.

# References

Amsden, Robert T., Howard E. Butler, and Davida M. Amsden. *SPC Simplified: Practical Steps to Quality.* New York: Quality Resources Press, 1998.

Barney, Matt. "Motorola's Second Generation," *Six Sigma Forum Magazine,* May 2002, 13–16.

Deming, W. Edwards. *Out of the Crisis.* Cambridge: Massachusetts Institute of Technology Center for Advanced Engineering Study, 1986.

Dew, John Robert. "Are You a Right-Brain or Left-Brain Thinker?" *Quality Progress,* April 1996, 91–93.

Folaron, Jim. "The Evolution of Six Sigma," *Six Sigma Forum Magazine,* August 2003, 38–44.

Gall, John. *Systemantics: The Underground Text of Systems Lore,* 2nd ed. Ann Arbor, MI: General Systemantics Press, 1986.

Gupta, Bhisham C. and H. Fred Walker. *Applied Statistics for the Six Sigma Green Belt.* Milwaukee: ASQ Quality Press, 2005.

Hare, Lynne. "SPC: From Chaos to Wiping the Floor," *Quality Progress,* July 2003, 58–63.

Harry, Mikel, and Richard Schroeder. *Six Sigma: The Breakthrough Management Strategy Revolutionizing the World's Top Corporations.* New York: Doubleday, 2000.

Harrington, James H., Glen D. Hoffherr, and Robert P. Reid. *Statistical Analysis Simplified: The Easy-to-Understand Guide to SPC and Data Analysis.* New York: McGraw-Hill, 1998.

Kaplan, Robert S., and David P. Norton. *The Balanced Scorecard.* Boston: Harvard Business School Press, 1996.

Langley, Gerald J., Kevin M. Nolan, Clifford L. Norman, Lloyd P. Provost, and Thomas W. Nolan. *The Improvement Guide: A Practical Approach to Enhancing Organizational Performance.* San Francisco: Jossey-Bass, 1996.

"Motorola Six Sigma Business Improvement Campaigns Brochure." Schaumburg, IL: Motorola University, 2003.

Roberts, Lon. *Process Reengineering: The Key to Achieving Breakthrough Success.* Milwaukee: ASQ Quality Press, 1994.

Zimmerman, Steven M., and Marjorie L. Icenogle, *Statistical Quality Control Using Excel,* 2nd ed. Milwaukee: ASQ Quality Press, 2003.

# Index

## A
accuracy, 24
Advocate General SPC case study, 17–22
  analysis of process, 18
  control limits, 21
  data collection plan, 20
  lessons of, 21–22
  management considerations, 21–22
  performance factors to monitor, 18–19
  process team members, 41–42
  process to monitor, 18
  SPC tools/techniques, 19–20
American Society for Quality (ASQ), 91
Amsden, Robert T., 66
assignable causes, 23
attribute charts, 72
attribute data, 63–64, 68–72
  defective items vs. individual defects, 69–70
attributes, 63
auditor, 81
average(s), 53–55, 58
average chart, 63–67
average moving range, 68
average value, 53–56, 60

## B
balanced scorecard, 2
baseline operating conditions, 74
baseline values, 26
bell-shaped curve, 59–60
best estimate, 37
brainstorming, 45–46, 93
breakthrough improvements, 88

## C
C-chart, 19, 63, 70–72
cause-and-effect diagram, 46
centerline value, 68, 73, 95
chart of individuals, 67
chart patterns, 26–34
charted measurements, 10
common causes, 23–24, 27, 29, 87
competitiveness, 24
consistency, 86
continuous improvement, 3, 24, 85–89
control chart, 6–7, 15, 41
  cumulative sum chart, 34–36
  cycling, 30–31
  mixture pattern, 32–33
  normal variation, 27
  out of control condition, 28–29
  pattern shift, 30

control chart (*continued*)
  scales and, 25
  selection of, 70, 72
  statistics and parameters of, 62
  stratification pattern, 32
  trending condition, 31
  understanding/interpreting of, 25
  unnatural patterns of, 29
  work-around solution, 29
control limits, 7, 14–16, 26, 32, 42, 64, 96
  defined, 9
  process capability and, 73–76
control points, 10, 42
corrective action, 31, 82–84
critical control points, 13
critical to quality (CTQ), 5
cross-functional task forces, 40
cumulative sum (CuSum) chart, 34–36
  example of, 36
  purpose of, 35
  stability/trending condition, 37
customer satisfaction, 2, 12, 39–40
customer's perception, 14
cycling pattern, 30–31

**D**

dashboard indicators, 2
data collection, 14, 77–84
  frequency of, 80–81
  inspection/review of results, 81–82
  responsible parties, 79–80
  where and how, 78
defective items, 69
defects per million opportunities (DPMO), 5
degree of spread, 64
desirable outcomes, 51
dispersion, 37, 57
distribution, 37–38

**E**

early warning indicators, 3
80/20 rule, 49
errors, 50

**F**

facilitator, 45
financial data, 2
firefighting, 23
fishbone diagram, 45–49, 93

**G**

go/no-go data, 69
grand average or grand mean, 64

**H**

Hare, Lynne, 6
human factor, 39–43

**I**

in-control process, 27
incremental improvements, 89
individual defects (flaws), 69
information flow diagram, 41
inspection, 3
intangible outputs, 11
interfaces, 48
internal processes, 2
Ishikawa, Kaoru, 46
Ishikawa diagram, 46

**K**

Kaplan, Robert, 2

**L**

learning and innovation, 2
lower control limit (LCL), 26, 53, 58, 65, 72

**M**

management by fact (MBF), 4, 6, 22
manufacturing processes, 1–2
mass production, 1, 39
mean, 53–56, 60

## Index  109

mean value, 55
measurable data, 63
measurement system
   charting individual values, 67–68
   monitoring attribute data, 68–72
   planning and charting of, 63–72
   X-bar and R-charts, 64–67
mixture pattern, 32–33
Motorola, 4–5
moving range (MR), 95

### N

natural systems, 60
non-value-adding activities, 42
normal distribution, 59–61, 86
   characteristics of, 37–38
normal operations, 29
normalcy, 68
Norton, David, 2
not-invented-here mentality, 39
NP-chart, 63, 70–72

### O

out-of-control process, 10–11, 15, 23, 28–29, 66

### P

P-chart, 19, 63, 70–72
parameters, 61–62
Pareto, Vilfredo, 49
Pareto analysis, 45, 49–51, 88, 93–94
Pareto principle, 49
pattern shift, 30
performance factors, 13–14
   early indicators, 14
   examples of, 13
   key metrics of, 2
   quantitative vs. qualitative measures, 14
population, 58–59
precision, 24

probability distribution, 61
process capability, 74
   control limits and, 73–76
process control, 1, 9
process flowcharts, 18, 42
   examples of, 43
process management, 29, 39–43
   customer satisfaction and, 39–40
   human factor and, 39–42
   mapping the process, 42–43
process orientation, 6
process owner, 40
process reengineering, 88
process team members, 40–41
product inspection, 3
proximity to problem source, 24

### Q

qualitative measures, 14
quality, 2, 12
   decentralizing responsibility for, 2
   options for dealing with, 3–4
   total quality management (TQM), 6
quality characteristics, 10
quantitative measures, 14

### R

R-bar, 65–66
R-chart, 20, 63–67, 71
random variation, 15
range, 55, 66
range values, 55
reliability, 24
representative sample, 59
run chart, 68

### S

S-chart, 67
sample, 9, 58–59
sample averages, 87
Shewhart, Walter, 6

sigma, 58
Six Sigma, 4–6, 58
Smith, Bill, 4–5
sources of errors, 50
special cause, 23, 27, 71
specification limits vs. control limits, 73–76
stability condition, 35
standard deviation, 26, 53, 56–58, 60
statistic, 61–62
statistical process control (SPC)
   Advocate General case (example), 17–22
   chart patterns, 26–34
   defined, 1
   environment of, 1–2
   framework for applying, 9–16
   human factors in, 22, 43
   management support, 6
   performance drivers, 2
   problem prevention orientation, 7
   problem solving by, 23–38
   process orientation, 6
   quality and, 3
   as quality management philosophy, 5–6
   special causes and, 24
   understanding/controlling variation, 7
   voice of the process data, 7
statistical process control (SPC) subsystem, 11
   analyze process, 12
   control limits, 14–16
   data collection plan, 14
   human judgment and, 16
   performance factors, 13–14
   process to monitor, 12
   setting up of, 11–16

SPC tool, 14
statistical quality control, 4
statistically valid, 68
statistics, 53–62
   averages and means, 53–55
   normal distribution, 60–61
   parameters and, 61–62
   populations and samples, 58–59
   range values, 55
   standard deviation, 56–58
stratification pattern, 32

**T**

tampered process, 33
tangible outputs, 11
total customer satisfaction, 39
total quality management (TQM), 6
trending condition, 31, 35

**U**

U-chart, 63, 70–72
undesirable outcomes, 51
unnatural patterns, 29
upper control limit (UCL), 26, 53, 58, 65, 72

**V**

value analysis, 88
variable data, 63
variables, 9, 63
variation, 23–24

**W**

Western Electric Manufacturing, 6
work-around solution, 29

**X**

X-bar, 20, 54, 56, 58, 71
X-bar chart, 63–67
X double-bar symbol, 65